EAI/Springer Innovations in Communication and Computing

Editor's Note

The impact of information technologies is creating a new world yet not fully understood. The extent and speed of economic, life style and social changes already perceived in everyday life is hard to estimate without understanding the technological driving forces behind it. This series presents contributed volumes featuring the latest research and development in the various information engineering technologies that play a key role in this process.

The range of topics, focusing primarily on communications and computing engineering include, but are not limited to, wireless networks; mobile communication; design and learning; gaming; interaction; e-health and pervasive healthcare; energy management; smart grids; internet of things; cognitive radio networks; computation; cloud computing; ubiquitous connectivity, and in mode general smart living, smart cities, Internet of Things and more. The series publishes a combination of expanded papers selected from hosted and sponsored European Alliance for Innovation (EAI) conferences that present cutting edge, global research as well as provide new perspectives on traditional related engineering fields. This content, complemented with open calls for contribution of book titles and individual chapters, together maintain Springer's and EAI's high standards of academic excellence. The audience for the books consists of researchers, industry professionals, advanced level students as well as practitioners in related fields of activity include information and communication specialists, security experts, economists, urban planners, doctors, and in general representatives in all those walks of life affected ad contributing to the information revolution.

Indexing: This series is indexed in Scopus, Ei Compendex, and zbMATH.

About EAI

EAI is a grassroots member organization initiated through cooperation between businesses, public, private and government organizations to address the global challenges of Europe's future competitiveness and link the European Research community with its counterparts around the globe. EAI reaches out to hundreds of thousands of individual subscribers on all continents and collaborates with an institutional member base including Fortune 500 companies, government organizations, and educational institutions, provide a free research and innovation platform.

Through its open free membership model EAI promotes a new research and innovation culture based on collaboration, connectivity and recognition of excellence by community.

More information about this series at http://www.springer.com/series/15427

Anandakumar Haldorai • Arulmurugan Ramu
Sudha Mohanram • Joan Lu
Editors

3rd EAI International Conference on Big Data Innovation for Sustainable Cognitive Computing

Springer

EAI
RESEARCH MEETS INNOVATION

Editors
Anandakumar Haldorai
Professor (Associate) and Research Head,
Department of Computer Science and Engineering
Sri Eshwar College of Engineering
Coimbatore, TN, India

Sudha Mohanram
Principal
Sri Eshwar College of Engineering
Coimbatore, TN, India

Arulmurugan Ramu
Professor, Department of Computer Science and Engineering
Presidency University
Bengaluru, KA, India

Joan Lu
Professor, Department of Computer Science, School of Computing and Engineering, Centre for Planning, Autonomy and Representation of Knowledge
University of Huddersfield
Huddersfield, UK

ISSN 2522-8595 ISSN 2522-8609 (electronic)
EAI/Springer Innovations in Communication and Computing
ISBN 978-3-030-78749-3 ISBN 978-3-030-78750-9 (eBook)
https://doi.org/10.1007/978-3-030-78750-9

This Springer imprint is published by the registered company Springer Nature Switzerland AG
The registered company address is: Gewerbestrasse 11, 6330 Cham, Switzerland

Preface

We are delighted to introduce the proceedings of the third edition of the 2020 European Alliance for Innovation (EAI) International Conference on Big Data Innovation for Sustainable Cognitive Computing (BDCC 2020). This conference has brought researchers, developers, and practitioners around the world who are leveraging and developing big data technology for a smarter and more resilient data. The theme of BDCC 2020 was "real-time ubiquitous data science."

The technical program of BDCC 2020 consisted of 17 full papers in oral presentation sessions at the main conference tracks. The conference tracks were: Main Track – Real-Time Ubiquitous Data Science and two workshop tracks being Track 1 – Workshop on Big Data in Intelligent and Information Systems and Track 2 – Workshop on Smart Sensor Networks and Big Data Management. Besides the high-quality technical paper presentations, the technical program also featured two keynote speeches designed for professionals working in the early stages of building an advancement program, as well as those with more mature operations. The two keynote speakers were Dr. Joan Lu, Professor, Department of Computer Science, School of Computing and Engineering, Centre for Planning, Autonomy and Representation of Knowledge University of Huddersfield, United Kingdom, and Dr. Manik Sharma, Department of Computer Science and Applications, DAV University, India.

Coordination with the steering chair Dr. Imrich Chlamtac and Dr. Anandakumar Haldorai was essential for the success of the conference. We sincerely appreciate Dr. Chlamtac's constant support and guidance. It was also a great pleasure to work with such an excellent organizing committee; we appreciate their hard work in organizing and supporting the conference. In particular, the technical program committee, led by our TPC chair Dr. Arulmurugan Ramu, publication chair Prof. M. Suriya, local committee chairs Prof. K. Karthikeyan and Prof. K. Aravindhan who have completed the peer-review process of technical papers and created a high-quality technical program. We are also grateful to our conference manager Ms. Karolina Marcinova and publication and managing editor Ms. Eliska Vlckova for their support and guidance. We thank all the authors who submitted and presented their papers at the BDCC 2020 conference and workshops.

We strongly believe that BDCC 2020 conference provides a good forum for all researcher, developers, and practitioners to discuss all science and technology aspects that are relevant to big data technology. We also expect that the future BDCC 2021 conference will be as successful and stimulating, as indicated by the contributions presented in this series.

Coimbatore, Tamil Nadu, India — Anandakumar Haldorai

Bengaluru, Karnataka, India — Arulmurugan Ramu

Coimbatore, Tamil Nadu, India — Sudha Mohanram

Huddersfield, UK — Joan Lu

Conference Organization

Steering Committee

Imrich Chlamtac Bruno Kessler Professor, University of Trento, Italy
Dr. Sudha Mohanram Sri Eshwar College of Engineering, Coimbatore, India
Dr. Anandakumar Haldorai Sri Eshwar College of Engineering, Coimbatore, Tamil Nadu, India
Dr. Arulmurugan Ramu Presidency University, Bangalore, India

Organizing Committee

General Chair

Dr. Anandakumar Haldorai Sri Eshwar College of Engineering, Coimbatore, India

TPC Chair

Dr. Arulmurugan Ramu Presidency University, Bangalore, India

Sponsorship and Exhibit Chair

Dr. Michaelraj Kingston Roberts Sri Eshwar College of Engineering, Coimbatore, India

Local Chair

Prof. Sivakumar K National Institute of Technology, Karnataka, India

Workshops Chair

Prof. Aravindhan K SNS College of Engineering, Coimbatore, India

Publicity & Social Media Chair

Dr. Akshaya V. S. Sri Eshwar College of Engineering, Coimbatore, India
Prof. Anand R Sona College of Technology, Salem, India

Publications Chair

Prof. Suriya Murugan KPR Institute of Technology, Coimbatore, India
Ms. Sudha B Vellore Institute of Technology, India

Web Chair

Prof. Karthikeyan K SNS College of Engineering, Coimbatore, India

Conference Manager

Ms. Karolina Marcinova EAI

Technical Program Committee

Dr. Chan Yun Yang National Taipei University, Taiwan.
Dr. Chow Chee Onn University of Malaya, Malaysia.
Dr. Sridevi Ravana University of Malaya, Malaysia.
Dr. Shahram Rahimi Southern Illinois University, Illinois, USA.
Dr. Marie Nathalie JAUFFRET Director of BECOM Program, Principality of Monaco.
Prof. Rojesh Dahal Kathmandu University, Kathmandu, Nepal.
Dr. Ram Kaji Budhathoki Nepal Engineering College, Nepal.
Dr. Hooman Samani National Taipei University, Taiwan.
Prof. M. D. Arif Anwary United International University, Bangladesh.
Prof. Fachrul Kurniawan Universitas Islam Negeri Maulana Malik Ibrahim Malang, Indonesia
Dr. Deepak B. Dhami Nepal Engineering College, Nepal.
Dr. K. R. Baskaran Kumaraguru College of Technology, Coimbatore, India
Dr. C. Malathy SRM Institute of Science and Technology, Tamil Nadu. India
Dr. G.K.D. Prasanna Venkatesan Karpagam University, Coimbatore, India.
Dr. M.G. Sumithra KPR Institute of Engineering and Technology, Tamil Nadu, India
Dr. U. Dinesh Acharya Manipal Institute of Technology, Manipal, India.
Dr. Latha Parameswaran Amrita Vishwa Vidyapeetham, Coimbatore, India
Dr. Bhabesh Nath Tezpur University Assam, India.
Dr. K.V. Prema, Professo Manipal Institute of Technology, Manipal, India.
Dr. E. Poovammal, Professor SRM Institute of Science and Technology, Tamil Nadu, India
Dr. Ashalatha Nayak Manipal Institute of Technology, Manipal, India.

Contents

About the Editors

Anandakumar Haldorai is associate professor and research head in the Department of Computer Science and Engineering, Sri Eshwar College of Engineering, Coimbatore, Tamil Nadu, India. He received his master's degree in software engineering and Ph.D. in information and communication engineering from PSG College of Technology affiliated to Anna University, Chennai. His research areas include big data, cognitive radio networks, mobile communications, and networking protocols. Dr. Haldorai has authored more than 125 research papers in reputed international journals and IEEE conferences. He has authored 11 books and many book chapters with reputed publishers such as Springer and IGI. He is editor-in-chief of KeAi–Elsevier *IJIN* and served as a reviewer for IEEE, IET, Springer, Inderscience, and Elsevier journals. Dr. Haldorai is also the guest editor of many journals with Elsevier, Springer, and Inderscience. He has been the general chair, session chair, and panelist in several conferences. He is senior member of IEEE, IET, and ACM and fellow member of EAI research group.

Arulmurugan Ramu is a professor at Presidency University, Bangalore, India. His research focuses on the automatic interpretation of images and related problems in machine learning and optimization. His main research interest is in vision, particularly high-level visual recognition. He has authored more than 35 papers in major computer vision and machine learning conferences and journals. Dr. Ramu is the recipient of Ph.D. degree in information and communication engineering from Anna University in Chennai, M.Tech. in information technology from Anna University of Technology, and B.Tech. degree in Information Technology. Dr. Ramu has guided several Ph.D. research scholars in the area of image processing using machine learning. He is an associate editor of Inderscience *IJISC*. He has been bestowed with Best Young Faculty Award 2018 and nominated for Best Young Researcher Award (Male) by the International Academic and Research Excellence Awards (IARE–2019).

Sudha Mohanram graduated from at the Government College of Engineering, Salem, and has obtained her master's degree in engineering from Coimbatore

Institute of Technology. She has completed her Ph.D. in electrical engineering at Anna University, Chennai, in 2010. She started her teaching profession as a lecturer in the Government College of Technology, Coimbatore. She possesses 20 years of teaching experience. When she was about 13 years into the teaching profession, her family founded Sri Eshwar College of Engineering in 2008. She has been playing the role of Secretary till 2011 and became the principal of the institution in 2011. Dr. Mohanram has steered the institution to be one of the most sought after institutions in Coimbatore within a short span of time through its laudable achievement in academic excellence and placement. She has published many papers in leading journals. She is a member of IEEE and EAI research group.

Joan Lu is professor in informatics at the University of Huddersfield (UK). Her extensive research covers information access, retrieval and visualization, XML and other data technologies, data management systems, security issues, wireless, and Internet computing. She has been an invited speaker at various industrial-oriented events; guest speaker in UK's Higher Education Academy (HEA) on mobile technology in classrooms, 2012; and keynote speaker for international conferences (INFOCCOM13, in Portugal and The Scientific Conference 2014, in Kurdistan). Professor Lu is a research grant reviewer for The Swiss National Science Foundation (SNSF), and an external adviser for a collaboration between Leeds Mat University, UK, and Sakarya University in Turkey. She was principal investigator for three EU projects from 2008 to 2012. She published 5 academic research books and over 180 papers. Professor Lu has acted as the founder and a program chair for the International XML Technology workshop and XMLTech (USA) for 11 years (2003–2011).

A Hybrid Algorithm for Document Clustering Using Optimized Kernel Matrix and Unsupervised Constraints

S. Siamala Devi, M. Deva Priya, P. Anitha Rajakumari, R. Kanmani, G. Poorani, S. Padmavathi, and G. Niveditha

Abstract Document clustering plays a dominant role in data mining, and grouping of data makes information retrieval easier. Significant information can be mined from a collection of documents by clustering them effectively. Several researches that concentrate on clustering documents are available in the literature. In the former works, document clustering is performed by using methodologies such as Term Weight-based Hybridized Harmony K-Means (TW-HHKM) and Coverage Factor-based Hybridized Harmony K-Means (CF-HHKM) searches. Clustering is normally performed using K-means algorithm, and cluster centroids are optimally found by using Harmony Search Algorithm (HSA). The main challenge faced by the existing methods is the reduced accuracy as unrelated documents may be grouped together. To overcome this problem, Novel Feature Weighting and Feature Selection-based Hybrid Scheme for Document Clustering (NFW-FS-HSDC) with optimized and unsupervised constraint kernel matrix K-means and Harmony Search Method (HSM) is introduced for accurate clustering of documents. The weights of the data instance and softness parameter decide the performance of spherical kernel K-means clustering. The parameter values for different datasets are identified after many trials. The kernel parameters are determined by applying Particle Swarm Optimization (PSO)-based technique. The proposed NFW-FS-HSDC algorithm

S. Siamala Devi · M. Deva Priya (✉) · G. Poorani · S. Padmavathi · G. Niveditha
Department of Computer Science and Engineering, Sri Krishna College of Technology, Coimbatore, Tamil Nadu, India
e-mail: s.siamaladevi@skct.edu.in; m.devapriya@skct.edu.in; poorani.g@skct.edu.in; padmavathi.s@skct.edu.in; niveditha.g@skct.edu.in

P. Anitha Rajakumari
Department of Computer Science and Engineering, SRM Institute of Science and Technology, Ghaziabad, Uttar Pradesh, India
e-mail: anitharp@srmist.edu.in

R. Kanmani
Department of Information Technology, Sri Krishna College of Engineering and Technology, Coimbatore, Tamil Nadu, India
e-mail: r.kanmani@skct.edu.in

A. Haldorai et al. (eds.), *3rd EAI International Conference on Big Data Innovation for Sustainable Cognitive Computing*, EAI/Springer Innovations in Communication and Computing, https://doi.org/10.1007/978-3-030-78750-9_1

optimizes the kernel matrix based on the feature ranges of the dataset by utilizing PSO. The exploratory tests are conducted on Newsgroup and Text REtrieval Conference (TREC) dataset, and it is obvious that the proposed NFW-FS-HSDC offers better Precision, Recall, F-Measure, Average Distance to Document Centroid (ADDC), Entropy, Overall Similarity and Cluster Purity.

Keywords Document clustering · Unsupervised constraints · Kernel matrix · Harmony Search method

1 Introduction

Data mining is a technique used for finding the interesting and hidden patterns from the given data. It plays a dominant role in all the fields, possibly to convert the data into valuable information [1]. It is generally used in a huge variety of profiling practices such as scientific discovery, fraud detection, surveillance and marketing. Data mining is also utilized to find out the patterns in a dataset but is frequently applied only on data samples. Generally, if the samples in the dataset are represented in a better way, then data mining becomes more effective. However, if the samples are not good, then the process becomes ineffective. As more number of unstructured web pages, commercial reports and scientific publications are increasingly available with the Information Retrieval (IR) systems and on the Internet, high-quality document clustering process becomes indispensable and significant in several applications such as web data management, data mining, search engine and even more.

1.1 Tasks of Data Mining

Traditional data mining tasks are divided into four classes:

- **Classification:** It organizes data into predefined groups. For instance, an email program may be classified as legitimate or spam. Commonly, algorithms such as Neural Networks (NN), Naive Bayesian Classification (NBC), Decision Tree Learning (DTL) and nearest neighbour methods are included for many real-time applications.
- **Clustering:** It is a procedure to find similar data and cluster them into groups using similarity measures. But, the groups are not predefined. So, various procedures and parameters are used to group similar data and identify outliers [2].
- **Regression:** It is a method for assessing the relationship among variables. It involves several mechanisms for demonstrating and examining variables with an emphasis on the connection between a variable and one or more independent

variables. This is a technique used to decrease error while organizing data using regression function [3].

- **Association Rule Mining (ARM):** This is a technique used to find the association among objects. For instance, a superstore may gather knowledge on client-buying practices. By applying ARM, the superstore can decide the products that are often bought together. This is often typically remarked as market basket analysis [4].

1.2 Clustering

Clustering is the procedure of considering the similarity that exists among the data to divide them and form groups with identical data objects. It deals with forming a group based on the likeness existing between data items. The data objects within the cluster have the highest similarity with huge deviation among clusters [5]. The inter-cluster and intra-cluster distances are measured using distinctive distance measures. Probably, the most primary variation between clustering and classification process is that clustering is an unsupervised learning system, whereas classification is a supervised one. For clustering, there is no need for a special mechanism to group the data items. Clustering is also known as automatic classification; but it is deceptive as it is fashioned without prior knowledge of classes. However, for classification, the classes of data are predefined. Clustering is a procedure that finds the cluster member based on the qualities and characteristics of data. On the other hand, in classification, classifiers find association among classes in the training dataset. In the dataset, data items are perfectly labelled by a specialist, and the trained classifier learns the relation among data and classes. Thus, the learned models are used to label data.

2 Related Work

Article clustering plays a dominant role in organizing documents in article publication organizations. The issues in arranging data are not clearly perceived and needs an expressive substance spotting of the produced clusters. To overcome the total inadequacy, an intensification of K-means algorithm referred to as Weighted-K-means (W-K-means) is proposed by Tran et al. [6]. W-K-means performs grouping by considering the broader meaning and then produces constructive labels for consequential groups. It offers 10 times better results when compared to the K-means algorithm by yielding more intra-cluster resemblance and less inter-cluster resemblance.

In partitioning-based document clustering algorithm, clustering results mainly depends on the number of clusters that is given as input to the algorithm. If the number of clusters is not known, then the results of clustering will be affected.

Agrawal and Phatak [7] have proposed a novel algorithm which generates quantity of clusters automatically for any unknown textual content dataset and clusters the records adequately based on the cosine similarity that exists between them. The system is capable of generating clusters for unknown data. But, when documents of the same category are to be clustered, there is a problem of over-clustering.

The fundamental task of document clustering is to evaluate the common similarities among the documents. The new method for calculating the likeness between documents is given by Jun et al. [8]. Many features are implanted in this symmetric measure. The dissimilarity between the presence and absence of features is essential rather than the variance that exists between the values related to a current feature. The similarity rises due to the decrease in the change among values. Moreover, the contribution of change is typically scaled. The likeness drops when the amount of presence or absence aspects increases. A feature that is not present does not contribute to the similarity. The propounded measure is applied to estimate the likeness among different units of documents. The measure is applied to a number of textual content applications together with single-labelled classification, multi-labelled classification, K-means and hierarchical agglomerative clustering for which the outcome acquired reveal the efficiency of the propounded similarity measure.

Every document dataset has its own features and choosing the right clustering algorithm that can manipulate all types of clusters is challenging. Clustering algorithms have their unique methods for finding the number of clusters. They may be hybridized to analyse the issues of single algorithms and increase performance. Sadeghian and Nezamabadi-pour [9] have propounded Gravitational Ensemble Clustering (GEC), a clustering approach that utilizes ensemble and gravitational clustering together for efficient grouping in contrast to the individual approaches such as K-Means algorithm and two hierarchical clustering methods namely, Single Link (SL) and Average Link (AL). The gravitational clustering based on Newton's gravitation law and ensemble clustering depends on two functions namely, generation and consensus function. These techniques offer better clustering accuracy, but the major shortcoming is the challenge faced in the optimization of continuous space.

Devi and Shanmugam [10] have proposed an effective ranking approach to rank the documents based on TF-IDF of the keywords in the documents. This reduces the communication overheads as needless documents are not downloaded. Furthermore, an effective query randomization scheme is propounded wherein multiple queries concerning the same search terms seem to be distinctive.

Fan et al. [11] have proposed a method that takes the advantages of Invasive Weed Optimization (IWO) algorithm, low complexity of K-means and hybridizing IWO with K-means. Different data experiments are carried out with various documents in different categories. The results show that IWO-K-means algorithm offers better clustering accuracy. The system finds optimal clusters but involves more convergence time.

Habibi and Popescu-Belis [12] have proposed a technique that considers an exact type of Just-In-Time (JIT) retrieval mechanism for conversational environments. This helps the users to identify the know-how requirements. Modelling the customer's know-how wishes by means of deriving implicit queries from brief dialog fragments is focused. These queries are established on sets of key terms extracted from the dialog. A clustering process is proposed to divide the set of keywords into smaller subsets independent of topics, thus constituting implicit queries. To improve performance of supervisory information clustering, a Semi-Supervised Concept Factorization (SSCF) is propounded by Lu et al. [13]. Pairwise constraints are incorporated into CF as reward and penalty terms. These terms assure that the data points in a cluster in the original space remain in the same cluster even in the transformed space.

Abualigah et al. [14] have propounded a scheme for feature selection called Feature Selection method by using PSO algorithm for Text Clustering (FSPSOTC) that generates a new subcategory of informative text features. This subgroup of features offers better performance involving less computational time. Investigations are performed for six standard text datasets with numerous features. These datasets are usually used in text clustering. The results reveal that this method enhances the efficiency of the text clustering scheme by dealing with a new category of informative features.

Handa et al. [15] have proposed a method that considers the association between keywords to cluster documents. The documents within the appropriate cluster are searched instead of the whole dataset.

2.1 Clustering Methods

The hybridization of K-means and HSM algorithm offers better clustering of document. Forsati et al. [16] have used TF-IDF as a feature for document clustering, that is, term-weighting scheme based on TF-IDF is used. In many documents, certain terms may appear just one or two times. Hence, the TF of words cannot bring out the actual words which lead to creation of tiny clusters in a document cluster. To overcome this, Coverage Factor (CF) feature is used in the proposed Novel Feature Weighting and Feature Selection-based Hybrid Scheme for Document Clustering (NFW-FS-HSDC).

Probably, K-means is the most widely used partitioning-based clustering algorithm which is appropriate for enormous datasets. It is self-effacing, simple, easy to implement and can be applied to a wide variety of applications. In this algorithm, the number of clusters must be specified in the preliminary step itself. It faces a challenge in producing local solution due to the selection of initial cluster centre randomly. Devi and Shanmugam [17] have proposed a Harmonic Search Method (HSM) which is a new meta-heuristic optimization procedure that emulates the tune improvisation process. It has been successful in handling the optimization crisis. In the hybridizing HSM with K-means, HSM is efficient in finding the best initial cluster centre for K-means to provide better results.

3 Hybridization of K-Means and Harmony Search Method (HSM)

K-Means algorithm when applied to document clustering always performs local search. The result or solution is always dependent on the previous step. The K-means algorithm uses arbitrarily selected data as centroid of every cluster and these values change in every iteration. The solution depends on the initial arbitrarily produced centroid values, whereas HSM is excellent in discovering the centroid values, but takes extra time to converge. As a way to overcome the above points of K-means and HSM, a hybrid algorithm that mixes both the ideas can offer effective outcome. Hybridization of K-means and HSM will also yield better results when compared to K-means algorithm [18].

The steps involved in HSM are detailed below.

Step 1: (i) Initialize the possible solutions as harmony and algorithm parameters.
(ii) Calculate the fitness function for document clustering using Average Distance of Documents to the Cluster Centroid (ADDC)
Step 2: Initialize the Harmony Memory (HM) with arbitrarily produced centroid using the fitness function of K-means
Step 3: Improve HM by adjusting the selected centroid
Step 4: If the adjusted centroid has better fitness, then update the HM
Step 5: If all the adjusted data of HM have same fitness value, then stop the search

Initially, the parameters of the HS algorithm which include Harmony Memory Measurements (HMMs), Harmony Memory Allowing Rate (HMAR), Pitch Tuning Rate (PTR) and the Quantity of Improvisations (QI) or the stopping criterion are unique. HMM is a reminiscence location wherein the solution vectors (sets of resolution variables) are saved. The HMAR and PTR are involved in the reinforcement of the solution vector. In HM initialization, the HM matrix is made with the arbitrarily produced resolution vectors similar to the HMM. Generating a new harmony is known as 'improvisation'.

3.1 Term Weight-Based Hybrid Scheme for Document Clustering (TW-HSDC)

In Term Weight-based Hybrid Scheme for Document Clustering (TW-HSDC), TF-IDF is used as a feature for document clustering. TF denotes the number of times a distinctive term appears in the document. Vector Space Model (VSM) is commonly utilized in document clustering. In the VSM, each report is represented using a feature vector. Traditionally, each characteristic resembles a key word or phrase that appears within the collection of documents. Every entry in the vector stores

Algorithm1: Hybridization of K-means and HSM

Input:

$D = \{d_1, d_2, \ldots \ldots d_n\}$ // set of n documents

$\{f_1, f_2, \ldots \ldots f_m\}$ - Features of a document

K - Number of clusters

HMM - Harmony Memory Measurement

HMAR - Harmony Memory Allowing Rate

PTR_{min} - Minimum Pitch Tuning Rate

PTR_{max} - Maximum Pitch Tuning Rate

MI - Maximum Number of Improvisation

NHM - New Harmony Memory

Output:

Clustered documents

// Find the initial centroid for clustering documents using HSM

// Cluster documents using K-means Clustering

HSM:

```
Initialize HM with random solutions
Assess the fitness of solutions in HM
Improvise NHM as follows:
for each i ∈[1,N] do
    if U(0, 1) ∈ HMAR then
        x_i = x^j_i where j ~ U (1,..., HMM)
        if(U(0, 1) ≤ PTR_t) then
            x_i = x_k^best , where best is the index of the harmony in HM
        else
            Perform arbitrary selection
        end if
    end if
end for
Assess the fitness of new solutions in NHM
Update HM with new solutions until maximum improvisation is attained
Choose a solution from the HM that has best fitness
Find the cluster centroids for the newly generated solution
```

K-means

Input:

```
D = {d_1, d_2, ... ... d_n} // Set of documents
C = {c_1, c_2, ... ... c_n} // Set of centroids
Calculate the distance between the data point and the cluster centers
for each data point 'd_i', find the closest centroid 'c_j' and assign 'd_i' to cluster 'c_j'
    Set ClusterID[i]=j; // j - ID of the closest cluster
    Set NearestDist[i]=d(d_i, c_j)
    for each cluster recalculate the centroid
      Repeat
        if each data point 'd_i' enters a cluster
          Compute the distance between the data points and the centroid of the closest cluster
          Place the data point in a cluster when the distance between the centroid of a cluster and data point is
          less than the distance between any other cluster center
        else
          for every centroid
              Estimate the distance d(d_i, c_j)
              Allocate 'd_i' to the cluster with adjoining centroid 'c_j'
              Set ClusterID[d_i]=j;
              Set NearestDist[d_i]= d(d_i, c_j);
          end for
        end if
        Re-compute the centroids
      Until there is no cluster updations
    end for
end for
```

the numerical weight for the respective feature of the document. The report is represented by using a vector of weights with 'n' elements.

$$d_i = (w_{i1}, w_{i2}, \ldots, w_{ij}, \ldots w_{in}) \tag{1}$$

where

w_{ij} – frequency of feature 'j' in document 'i'
n – number of features

It is a term-weighting scheme designed for know-how retrieval (as ranking in search engines like Google) IS used to find the right use of features in record classification and clustering. Weight of a characteristic can be found using the ensuing formula

$$w_{ij} = \mathrm{TF} - \mathrm{IDF} = \mathrm{TF}\,(i, j)\,.\log\frac{N}{df(j)} \tag{2}$$

where

TF(i, j) – frequency of feature 'j' in a document 'd_i'
N – number of documents in the collection
$df(j)$ – number of records in which feature 'j' appears

The algorithm TW-HSDC uses TF of features to cluster the documents. As a result, only low-level clusters are formed. The TF-IDF model does not consider words which appear once or twice in the document. There are chances for the words to be used for defining the category of documents. The significance of these words cannot be identified in TF.

3.2 Coverage Factor-Based Hybrid Scheme for Document Clustering (CF-HSDC)

In Coverage Factor-based Hybrid Scheme for Document Clustering (CF-HSDC), the hybridization of K-means and HSM uses the Coverage Factor (CF) as a feature for document clustering. The drawbacks of the TF-based document clustering are eliminated by using the CF. The coverage of the features is outlined as the percentage of files containing a minimum possibility of an important term.

The CF feature considers the significance of terms. CF-HSDC calculates the document frequency of words and performs word selection based on the threshold values. The CF focusses on the importance of less frequent terms by considering the percentage of documents in the dataset containing these words. The coverage of terms is calculated based on the number of input documents taken for clustering and the number of clusters. Based on the number of clusters and percentage of term availability, the lower and upper ranges of coverage are automatically decided.

Algorithm 2: Coverage Factor

Step 1: Arbitrarily choose a subgroup of documents of size 'm' from the whole set of documents
Step 2: Mine the terms that appear atleast once in the documents
Step 3: Eradicate stop words and join words with the same root using stemming
Step 4: Find the Document Frequency (DF) of words that are mined in the former step
Step 5: Set lower=K and upper=K
Step 6: Choose words with DF that are in the range 'lower' to 'upper'
Step 7: If the word coverage is more than the pre-defined threshold, then Stop. Else Set lower=lower-1 and upper=upper+1 and goto previous step

Here 'K' is the cluster size. The number of clusters is almost equal to 'N/K'. As the document set is typically huge, it is ineffective to perform feature extraction on the whole set of documents. To extract good features, a set of sample documents is randomly selected. A stop word list is generated which includes the meaningless words to be removed from the documents. Stemming is applied to combine similar words (Example: Learning, Learned, Learns can be represented as Learn). Since feature reduction leads to a decrease of clustering time, steps 4 to 6 try to minimalize the number of features and get sensible attention for the features. For example, the user wants the resultant cluster to comprise of 'K' documents. Initially, features with DF equal to 'K' (Step 4) are chosen. The range {lower, upper} is increased recurrently in step 6. This is to ensure sufficient coverage for the resulting feature set. The exact number of clusters is not equal to 'N/K'. The proposed method makes use of a coverage threshold to guarantee that the chosen feature has adequate coverage of all important terms that occur less frequently.

4 Novel Feature Weighting and Feature Selection-Based Hybrid Scheme for Document Clustering (NFW-FS-HSDC)

The proposed Novel Feature Weighting and Feature Selection-based Hybrid Scheme for Document Clustering (NFW-FS-HSDC) includes the following steps.

4.1 Optimized Kernel Matrix K-Means Algorithm

In this method, the parameters including weight, softness, number of clusters of the general kernel function are tuned using PSO to improve discrimination in the higher dimensional space. Thus, optimized kernel is formed for improving the accuracy of document clustering. In this method, PSO is used for finding the optimized kernel matrix by identifying optimal weight and the membership values of each object. In every iteration, each particle has a set of kernel matrices that are derived from data objects along with the updated weight and membership matrix. The optimized kernels are very useful for calculating the similarity among heterogeneous features. An objective function '*J(P)*' for fine-tuning the parameters of vector '*P*' is defined using PSO:

$$J(P) = \frac{\sum_{i=1}^{c} \sum_{j=1}^{n_i} d\,(j, i, i)}{\sum_{i=1}^{c} \sum_{j=1, j \neq i}^{c} \sum_{k=1}^{n_i} d\,(k, i, j)} \tag{3}$$

$d(j, i, i)$ – Distance between successive *i*th vector in *j*th cluster
$d(k, i, j)$ – Distance between *i*th vector in *j*th cluster in total number of clusters

The distance between the *i*th vector in the *j*th cluster (u_{ij}) with the centroid of the *q*th cluster in the higher dimensional space is given by

$$d\,(i, j, q) = k\left(u_{ij}, u_{ij}\right) - \left(\frac{2}{n_q}\right) \sum_{k=1}^{n_q} k\left(u_{ij}, u_{kq}\right) + \left(\frac{1}{n_q^2}\right) \sum_{k=1}^{n_q} \sum_{l=1}^{n_q} k\left(u_{kq}, u_{lq}\right) \tag{4}$$

$d(i, j, q)$ – Distance between the *i*th vector in the *j*th cluster (u_{ij}) with the centroid of the *q*th cluster
$k(u_{ij}, u_{ij})$ – *i*th vector in *j*th cluster
u_{ij} – *i*th vector in *j*th cluster
n_q – Centroid of cluster '*q*'
$k(u_{ij}, u_{kq})$ – *i*th vector in *j*th cluster; *k*th vector in *q*th cluster
n_q^2 – Centroid of cluster '*q*'
$k(u_{kq}, u_{lq})$ – *k*th vector in *q*th cluster; *l*th vector in *q*th cluster

Algorithm 3: PSO for optimizing Vector 'P'

```
J(R) - Objective function of Random Vector
R() - Random Vector
p_i - i^th particle
 Initialize the number of the particles (n) and the number of iterations (N)
 Generate the random Vector 'R' with size 25*1
 if J(R)>0 and R(21:24)>0
     Assign p_i =R and i=i+1
 else
     Generate 'R' of size 25*1
 if i<n
     Generate 'R' of size 25*1
 else
     Generate a valid parameter
```

Thus, optimization maximizes the discrimination in the higher dimensional space and increases the accuracy of the clustering process.

4.2 *Unsupervised Constraint Kernel Matrix K-Means*

Unsupervised constraint clustering is performed by the machine itself. It includes two types of constraints, namely, word and document constraints.

- Word constraints depend on the word classifications learned from auxiliary corpus. The must-links are added if two words are close to each other semantically. Named entity identification and overlapped named entities are used to frame constraints for documents. The constraints are obtained by calculating the semantic distance from synonyms in WordNet.
- In case of document constraints, the must-link constraints are built from the connected entities such as person, location, organization, etc.

5 Results and Discussion

The proposed NFW-FS-HSDC offers better Precision, Recall, F-Measure, Average Distance to Document Centroid (ADDC), Entropy, Overall Similarity and Cluster Purity for Newsgroup and Text REtrieval Conference (TREC) datasets. From these representations, it is evident that the propounded scheme outperforms the existing methodologies in fine-tuned clusters. Hence, the effort and price of using physically labelled constraints are reduced by implementing the unsupervised constraints.

5.1 *Performance of the Proposed NFW-FS-HSDC Scheme for Newsgroup Dataset*

The clustering results of existing and proposed approaches for Newsgroup dataset are clearly given in Table 1.

For Newsgroup dataset, the proposed NFW-FS-HSDC scheme offers 5.6% and 4.2% better Precision in contrast to the existing TF-HSDC and CF-HSDC schemes. Similarly, the propounded NFW-FS-HSDC scheme offers 5% and 3.7% better Recall when compared to the existing TF-HSDC and CF-HSDC schemes.

Table 1 Results of document clustering for Newsgroup dataset

	Methods		
Parameters	TF-HSDC	CF-HSDC	NFW-FS-HSDC
Precision	90.05	91.23	95.07
Recall	90.39	91.53	94.94
F-Measure	90.22	91.38	95.01
ADDC	73.17	66.73	60.34
Entropy	86.6	84.7	79.33
Overall Similarity	84.37	85.21	92.81
Cluster Purity	86.63	87.92	94.46

Table 2 Results of document clustering for TREC dataset

	Methods		
Parameters	TF-HSDC	CF-HSDC	NFW-FS-HSDC
Precision	88.93	90.41	93.98
Recall	89.27	89.89	94
F-Measure	89.1	90.15	93.99
ADDC	70.81	64.33	57.7
Entropy	82.82	81.03	77.99
Overall Similarity	81.81	86.39	89.88
Cluster Purity	86.74	88.17	92.07

The benchmarked TF-HSDC and CF-HSDC schemes offer 5.3% and 4% reduced F-Measure in contrast to the proposed NFW-FS-HSDC scheme. Similarly, the benchmarked TF-HSDC and CF-HSDC schemes offer 21.3% and 10.6% reduced ADDC when compared to the proposed NFW-FS-HSDC scheme.

The proposed NFW-FS-HSDC scheme offers 9.2% and 6.8% better Entropy in contrast to the existing TF-HSDC and CF-HSDC schemes. The proposed NFW-FS-HSDC scheme offers 10% and 8.9% improved Overall Similarity in contrast to the existing TF-HSDC and CF-HSDC schemes. Similarly, the benchmarked TF-HSDC and CF-HSDC schemes offer 9% and 7.4% reduced Cluster Purity when compared to the proposed NFW-FS-HSDC scheme.

5.2 *Performance of the Proposed NFW-FS-HSDC Scheme for TREC Dataset*

Similarly, the results of existing and proposed approaches for TREC dataset are shown in Table 2.

For TREC dataset, the proposed NFW-FS-HSDC scheme offers 5.7% and 3.9% better Precision in contrast to the existing TF-HSDC and CF-HSDC schemes.

Similarly, the propounded NFW-FS-HSDC scheme offers 5.3% and 4.6% better Recall when compared to the existing TF-HSDC and CF-HSDC schemes. The benchmarked TF-HSDC and CF-HSDC schemes offer 5.5% and 4.3% reduced F-Measure in contrast to the proposed NFW-FS-HSDC scheme. Similarly, the benchmarked TF-HSDC and CF-HSDC schemes offer 22.7% and 11.5% reduced ADDC when compared to the proposed NFW-FS-HSDC scheme.

The proposed NFW-FS-HSDC scheme offers 6.2% and 3.9% better Entropy in contrast to the existing TF-HSDC and CF-HSDC schemes. The proposed NFW-FS-HSDC scheme offers 9.9% and 4% improved Overall Similarity in contrast to the existing TF-HSDC and CF-HSDC schemes. Similarly, the benchmarked TF-HSDC and CF-HSDC schemes offer 6.1% and 4.4% reduced Cluster Purity when compared to the proposed NFW-FS-HSDC scheme.

It is seen that Precision (Fig. 1), Recall (Fig. 2), F-Measure (Fig. 3), ADDC (Fig. 4), Entropy (Fig. 5), Overall Similarity (Fig. 6) and Cluster purity (Fig. 7) for Newsgroup and TREC datasets of the proposed NFW-FS-HSDC scheme are better in contrast to the benchmarked TF-HSDC and CF-HSDC schemes considered for investigation.

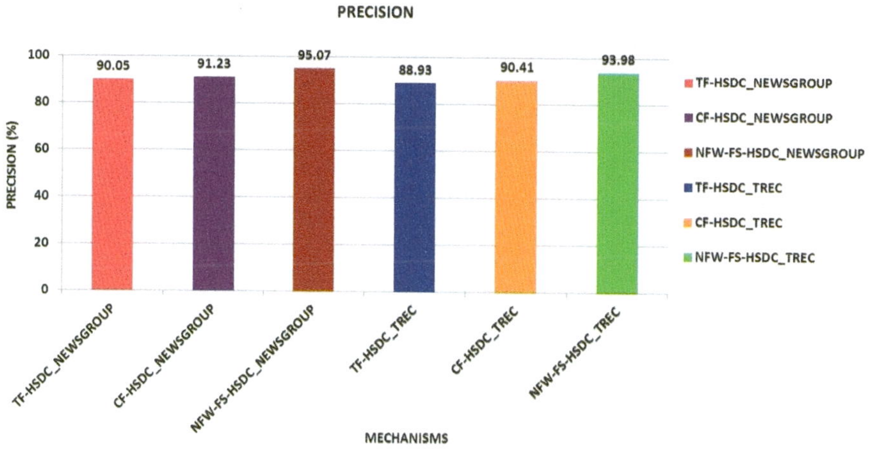

Fig. 1 Precision of the TF-HSDC, CF-HSDC and NFW-FS-HSDC schemes for the Newsgroup and TREC dataset

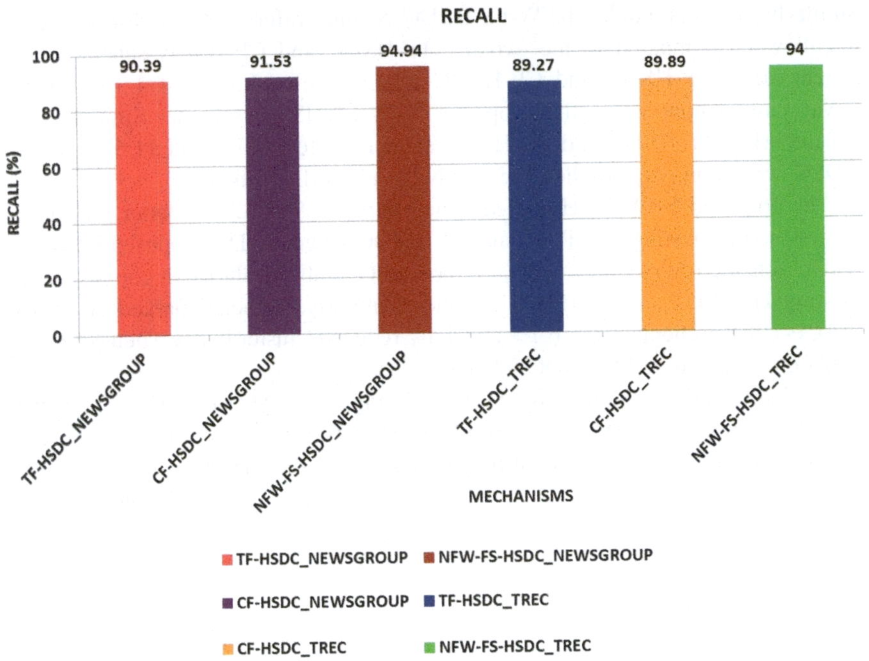

Fig. 2 Recall of the TF-HSDC, CF-HSDC and NFW-FS-HSDC schemes for the Newsgroup and TREC dataset

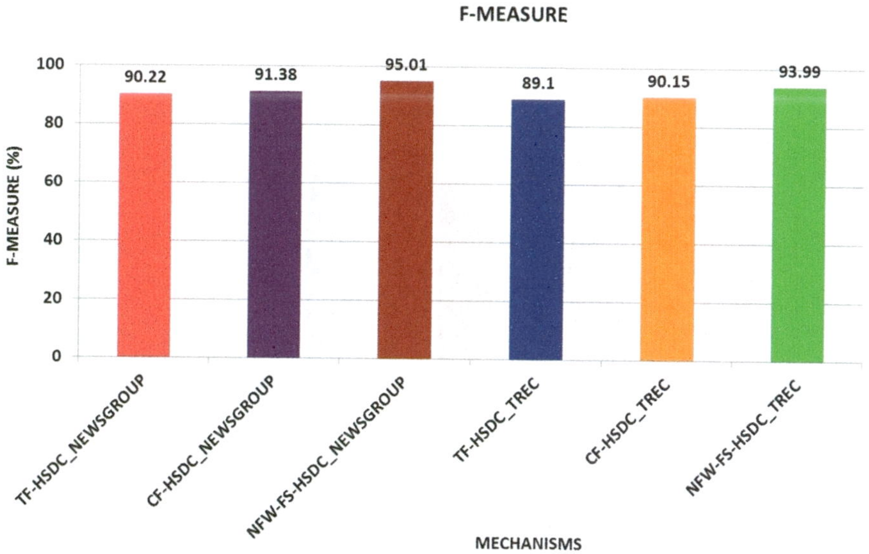

Fig. 3 F-Measure of the TF-HSDC, CF-HSDC and NFW-FS-HSDC schemes for the Newsgroup and TREC dataset

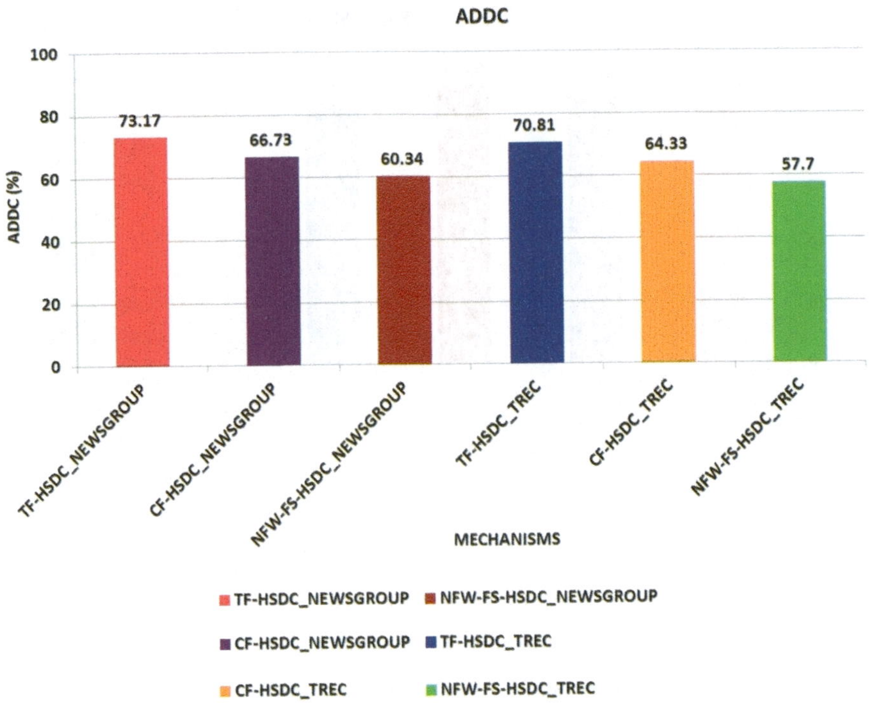

Fig. 4 ADDC of the TF-HSDC, CF-HSDC and NFW-FS-HSDC schemes for the Newsgroup and TREC dataset

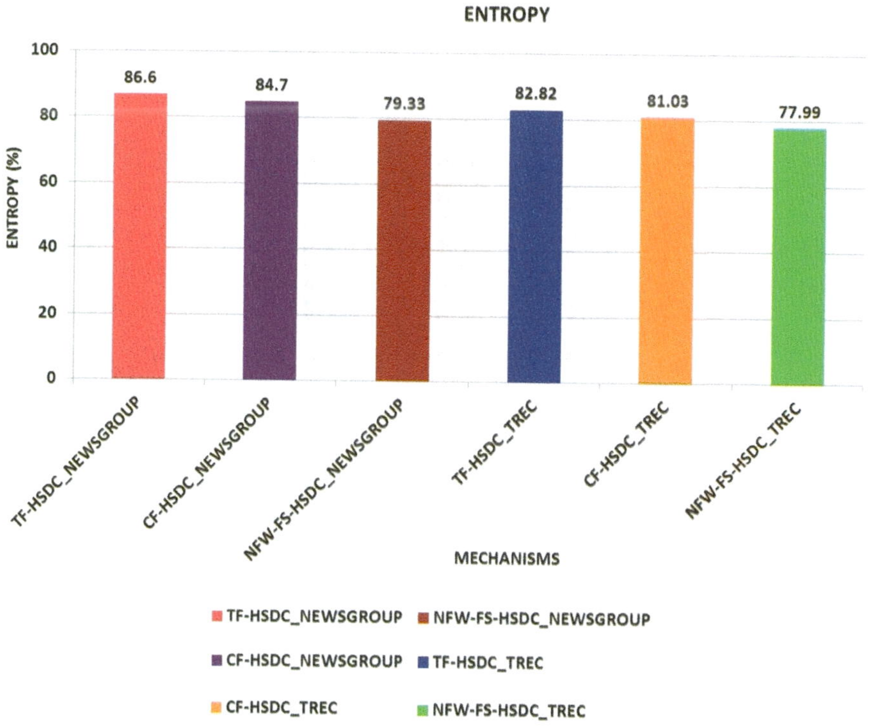

Fig. 5 Entropy of the TF-HSDC, CF-HSDC and NFW-FS-HSDC schemes for the Newsgroup and TREC dataset

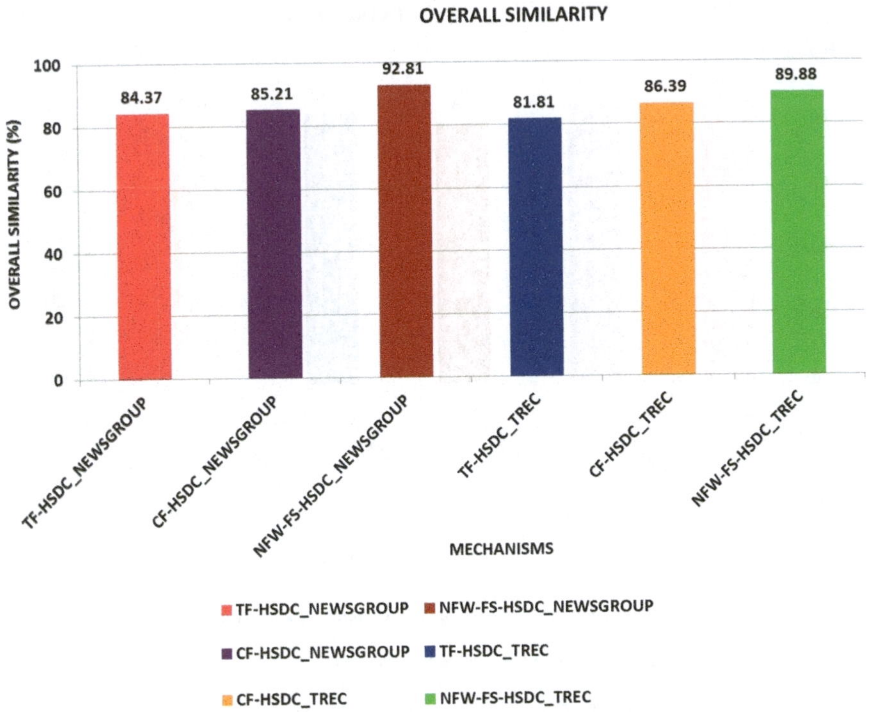

Fig. 6 Overall Similarity of the TF-HSDC, CF-HSDC and NFW-FS-HSDC schemes for the Newsgroup and TREC dataset

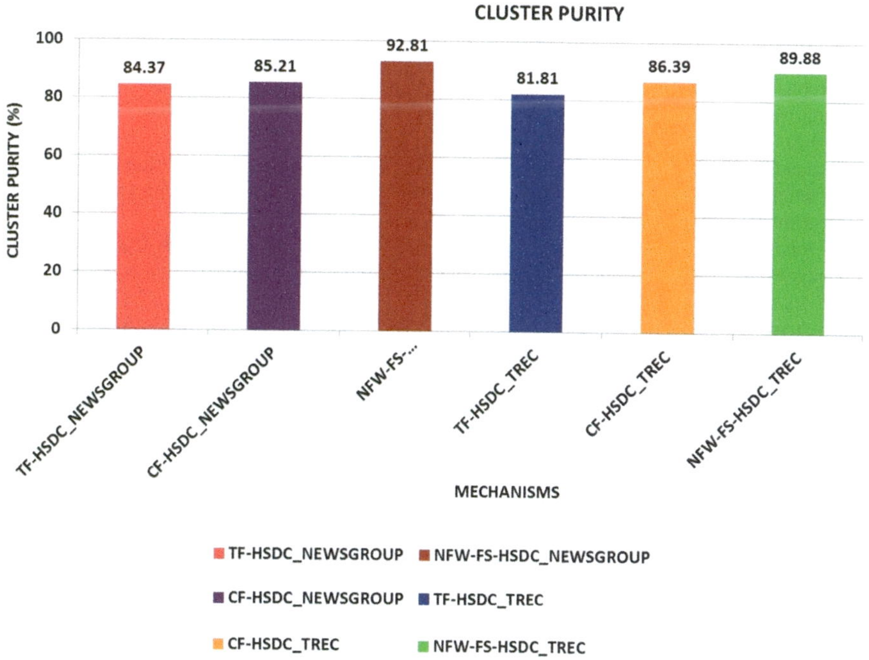

Fig. 7 Cluster Purity of the TF-HSDC, CF-HSDC and NFW-FS-HSDC schemes for the Newsgroup and TREC dataset

6 Conclusion

Different techniques such as Term Frequency-based Hybrid Scheme for Document Clustering (TF-HSDC), Coverage Factor-based Hybrid Scheme for Document Clustering (CF-HSDC) and Novel Feature Weighting and Feature Selection-based Hybrid Scheme for Document Clustering (NFW-FS-HSDC) are proposed in this Chapter to facilitate efficient document clustering. It is important to observe that the K-means algorithm has the limitation of generating local optimal solution, whereas Harmony Search Method (HSM) takes more time to converge. In the TF-HSDC algorithm, clusters are produced based on only TF features count. In CF-HSDC, CF does not guarantee that all documents are considered for the clustering process. In the proposed NFW-FS-HSDC, a single viewpoint is used to cluster documents. The effectiveness of single viewpoint clustering mainly depends on the appropriateness of the similarity measure. The experimental results prove that the NFW-FS-HSDC outperforms the other two clustering methods in terms of Precision, Recall, F-Measure, ADDC, Entropy, Overall Similarity and Cluster Purity for Newsgroup and TREC datasets.

References

1. Rubiano, S. M. M., & Garcia, J. A. D. (2016). Analysis of data mining techniques for constructing a predictive model for academic performance. *IEEE Latin America Transactions, 14*(6), 2783–2788.
2. Aouad, L. M., Le-Khac, N. A., & Kechadi, T. M. (2007). Lightweight clustering technique for distributed data mining applications. In *Industrial conference on data mining* (pp. 120–134). Berlin/Heidelberg: Springer.
3. Gaobo, C., & Xiufang, C. (2011). Combining partial least squares regression and least squares support vector machine for data mining. In *IEEE international conference on E-business and E-government* (pp. 1–4).
4. Bouziri, A., Latiri, C., Gaussier, E., & Belhareth, Y. (2015). Learning query expansion from association rules between terms. In *7th international joint conference on knowledge discovery, knowledge engineering and knowledge management* (Vol. 1, pp. 525–530).
5. Kumar, N., Verma, V., & Saxena, V. (2013). Cluster analysis in data mining using k-means method. *International Journal of Computer Applications, 76*(12), 11–14.
6. Tran, T. N., Wehrens, R., & Buydens, L. M. (2006). KNN-kernel density-based clustering for high-dimensional multivariate data. *Computational Statistics & Data Analysis, 51*(2), 513–525.
7. Agrawal, R., & Phatak, M. (2013). A novel algorithm for automatic document clustering. In *3rd IEEE international advance computing conference* (pp. 877–882).
8. Jun, S., Park, S. S., & Jang, D. S. (2014). Document clustering method using dimension reduction and support vector clustering to overcome sparseness. *Expert Systems with Applications, 41*(7), 3204–3212.
9. Sadeghian, A. H., & Nezamabadi-pour, H. (2015). Document clustering using gravitational ensemble clustering. In *International symposium on IEEE artificial intelligence and signal processing* (pp. 240–245).
10. Devi, S. S., & Shanmugam, A. (2015). An integrated harmony search method for text clustering using a constraint based approach. *Indian Journal of Science and Technology, 8*(29), 1–7.
11. Fan, C., Zhang, T., Yang, Z., & Wang, L. (2015, August). A text clustering algorithm hybriding invasive weed optimization with K-means. In *12th IEEE international conference on ubiquitous intelligence and computing and 12th IEEE international conference on autonomic and trusted computing and 15th IEEE international conference on scalable computing and communications and its associated workshops* (pp. 1333–1338).
12. Habibi, M., & Popescu-Belis, A. (2015). Keyword extraction and clustering for document recommendation in conversations. *IEEE/ACM Transactions on Audio, Speech, and Language Processing, 23*(4), 746–759.
13. Lu, M., Zhao, X. J., Zhang, L., & Li, F. Z. (2016). Semi-supervised concept factorization for document clustering. *Information Sciences, 331*, 86–98.
14. Abualigah, L. M., Khader, A. T., & Hanandeh, E. S. (2018). A new feature selection method to improve the document clustering using particle swarm optimization algorithm. *Journal of Computational Science, 25*, 456–466.
15. Handa, R., Krishna, C. R., & Aggarwal, N. (2019). Document clustering for efficient and secure information retrieval from cloud. *Concurrency and Computation: Practice and Experience, 31*(15), e5127.
16. Forsati, R., Meybodi, M., Mahdavi, M., & Neiat, A. (2008). Hybridization of k-means and harmony search methods for web page clustering. *In: IEEE/WIC/ACM International Conference on Web Intelligence and Intelligent Agent Technology, 1*, 329–335.
17. Devi, S. S., & Shanmugam, A. (2014). Hybridization of K-means and harmony search method for text clustering using concept factorization. *International Journal of Advanced Research in Computer Engineering & Technology, 3*(8), 2685–2689.
18. Devi, S. S., & Shanmugam, A. (2016). Hybridized harmony search method for text clustering using concept factorization. *International Journal of Advanced Computer Technology, 2016*, 320–327.

A Multi-objective Optimal Trajectory Planning for Autonomous Vehicles Using Dragonfly Algorithm

R. Syama and C. Mala

Abstract Trajectory planning is considered as a major challenge in autonomous driving, which faces significant issues like safety and efficiency. This chapter proposes a novel swarm intelligence meta-heuristic optimization algorithm, called dragonfly algorithm, for the lane-change behaviour in the navigation of autonomous vehicles. A multi-objective lane-changing trajectory planning method has been proposed to optimize the trajectory and avoid collision, which mimics the dynamic and static swarm behaviours of the natural dragonflies. Whenever the autonomous vehicle senses an obstacle, automatically the lane-change manoeuvre should take place. The feasibility and the effectiveness of the algorithm are verified by simulation results using the lane-change data from the benchmark NGSIM dataset. Simulation results show that the proposed algorithm for trajectory planning gives an optimal path for lane-change scenario considering both static and dynamic obstacles.

Keywords Autonomous driving · Trajectory planning · Dragonfly algorithm · Lane-change manoeuvre

1 Introduction

Autonomous driving refers to self-driving vehicles that sense the environment and make decisions by their own without the involvement of a human driver. Autonomous driving has been one of the keen areas of research for the past few decades due to its ability to enhance the safety and efficiency of transportation system.

R. Syama (✉) · C. Mala
Department of Computer Science and Engineering, National Institute of Technology, Tiruchirappalli, Tamil Nadu, India
e-mail: 406119007@nitt.edu; mala@nitt.edu

A. Haldorai et al. (eds.), *3rd EAI International Conference on Big Data Innovation for Sustainable Cognitive Computing*, EAI/Springer Innovations in Communication and Computing, https://doi.org/10.1007/978-3-030-78750-9_2

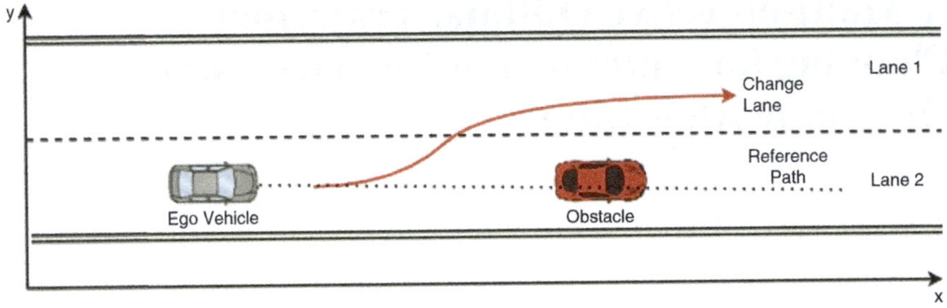

Fig. 1 Lane-change scenario

Trajectory planning has a critical role in autonomous driving technology, and it is one of the major challenges that must be addressed. In the real-world scenario, the autonomous vehicles must navigate autonomously and take intelligent decisions. The vehicle must determine a path for travel from the data it collected from the static and dynamic environments. Then, the trajectory planning is done to obtain an optimal path for the smooth movement from the source to destination achieving certain constraints. It is really important and hard to generate a path that is optimal at the same time meet the real-time requirements. Path planning involves searching and computation of optimal collision-free paths. The feasible path of the autonomous vehicle is generated considering the vehicle geometry, its surroundings, kinematic controls, etc.

The central idea of this chapter is to design a trajectory planning technique that can automatically generate an optimal path from the start to the goal position. Figure 1 shows the lane-changing process by autonomous vehicle, say ego vehicle, while avoiding obstacle. The obstacle avoidance trajectory must consider the safety criteria, mainly the longitudinal and lateral distances with the obstacles. To generate the optimal path, this chapter proposes a meta-heuristic Dragonfly Algorithm (DA). It is based on swarming behaviours of the natural dragonflies [1, 2]. Several researches have been done in trajectory planning for autonomous vehicles, and most of them are based on the searching techniques [3, 4]. But, these techniques have failed in considering obstacle avoidance behaviour.

2 Related Works

The lane-changing behaviour is essential to perform different driving activities such as road merging, entry and exit to a highway, vehicle overtaking, etc. [2]. Several studies have been done to meet the trajectory planning problem for the lane changing in different areas. The major challenge in generating a trajectory is to ensure the feasibility while considering the different constraints. Different optimization algorithms have been applied to trajectory planning problem in recent years. The commonly used algorithms are Ant Colony Optimization [5], Genetic Algorithms [6], Particle Swarm Optimization [7], etc.

In global planning approach, a path is generated from the initial position to the end position from the prior information about the environment. Several methods have been used for global planning such as A* algorithm [8], RRT algorithm [9] and Dijkstra's algorithm [10]. Major drawback of these algorithms is that the whole trajectory calculation process is very time consuming. Therefore, these algorithms are not suitable for trajectory planning of real-time applications such as autonomous vehicles where the obstacles are mainly dynamic.

Local trajectory planning techniques are widely used in most of the real-time applications. These approaches calculate the trajectories for a limited time window, which considers the environment conditions. Several geometric algorithms are used for the local trajectory planning such as splines [11], Bezier curves [12], Clothoid curves [13, 14], etc. These methods generate smooth trajectories by connecting the waypoints for the navigation of the vehicles. But, these methods also suffer from long computation time. Sigmoid curve–based approach is used in [15] for local trajectory planning for obstacle avoidance behaviour. The safe space between the autonomous vehicle and the obstacles is also taken into consideration.

A collision-free trajectory planning method using B spline and RRT-based method is used to improve the robustness of motion planning in autonomous vehicles [16]. Still there exists a timeout possibility. A non-linear Model Predictive Control approach for vehicle navigation is suggested, which considers collision-free trajectory [13]. It generates a dynamic obstacle avoiding trajectory planning with certain constraints. It failed to study the random movement of moving obstacles. The real-time motion planner is proposed in [17, 18]. A number of vision-based approaches are also applied to path planning problems [19, 20].

In this chapter, Dragonfly algorithm is proposed for optimizing the vehicle trajectory. It is a new approach that has powerful capabilities to solve different optimization problems. Studies have shown that DA performs better compared to PSO and gains several multimodal test functions [21]. This approach has several advantages, which makes it suitable for real-world applications:

1. Convergence is guaranteed during optimization since the weights are changed adaptively.
2. It always converges to the global optimum.
3. Randomness can be added to the algorithm.

3 Proposed Method

This chapter proposes a trajectory planning method that generates an optimal collision avoidance trajectory for the autonomous vehicle using Dragonfly Algorithm (DA). Dragonfly algorithm is an optimization technique, which was proposed by Seyedali Mirjalili in 2015. It is based on the static and dynamic grouping behaviour

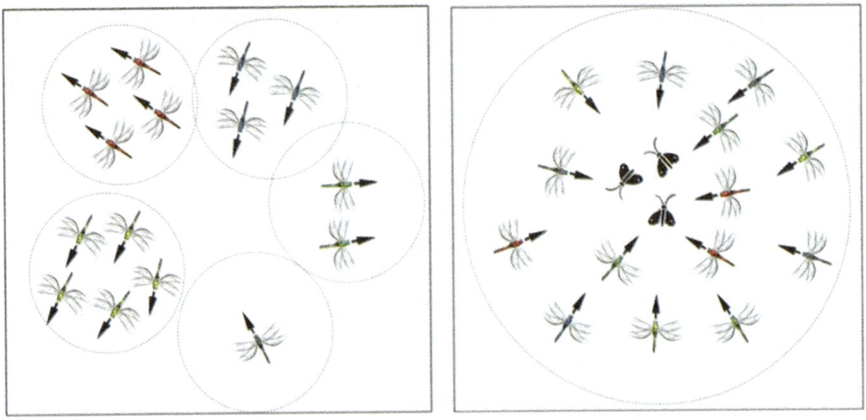

Fig. 2 Dynamic and static dragonfly swarms

of biological dragonflies, which mimics the different phases of optimization and can be employed to solve a wide range of optimization problems [21]. Dragonflies are small insects that hunt and eat other small creatures like butterflies, bees etc. [22]. Dragonflies have a unique swarming behaviour. They form groups only for two reasons: hunting and migration [23]. The swarm formed for hunting is known as static (feeding) swarm and the swarm formed for relocation is known as dynamic swarm. The static and dynamic swarms [21] are illustrated in Fig. 2.

In static swarm, small groups of dragonflies move forward and backward to hunt other flying insects. A swarm is dynamic where a huge number of dragonflies are grouped to move in a single direction over a long distance. These dynamic and static swarms compose the exploitation and exploration stages of DA. These two behaviours are in accordance with the meta-heuristic stages [21].

The dragonfly individuals exhibit five primitive behaviours which can be used to model the swarm behaviour. The dragonflies exhibit five properties, which are shown in Fig. 3 [21]. Each of the behaviour of the dragon flies are modelled as follows [21]:

1. *Separation (Si)* represents the operation to avoid collisions that the individuals follow with other individuals in the region. It is calculated as [21]:

$$Si = -\sum_{j=1}^{N} X - Xj \tag{1}$$

where X is the location of the dragonfly, Xj is the location of jth nearest individual, and N is the number of nearby individuals.

2. *Alignment* (*Ai*) represents the dragonfly's velocity which matches with the other nearby dragonflies of the same group. It is calculated as [21]:

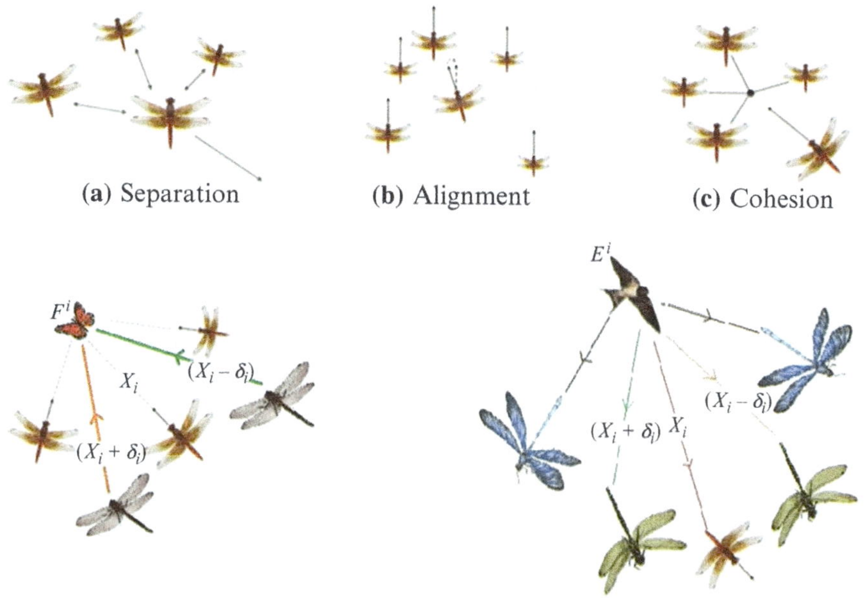

(a) Separation **(b)** Alignment **(c)** Cohesion

(d) Modified steps for Attraction towards the food **(e)** Modified steps for Distraction from the enemy

Fig. 3 Behaviour of dragonflies

$$Ai = \frac{\sum_{j=1}^{N} Vj}{N} \tag{2}$$

where V_j represents the velocity of the jth dragonfly.

3. Cohesion (*Ci*) is the tendency of the individual to move towards the centre of the group. It can be calculated as [5]:

$$Ci = \frac{\sum_{j=1}^{N} Xj}{N} - X \tag{3}$$

4. Attraction (*Fi*) represents the attraction towards food source and can be represented as [5] :

$$Fi = X^{+} - X \tag{4}$$

where F_i represents the food source of the ith individual.

5. Distraction (*Ei*) is the distraction from enemies and is represented as [5]:

$$Ei = X^{-} - X \tag{5}$$

where E_i denotes the location of the enemy of the ith dragonfly

The step vector ΔX and the position X can be used to update the location of the dragonflies in the search domain. The update is done as follows [21]:

$$\Delta X_i^{t+1} = (sS_i + aA_i + cC_i + fF_i + eE_i) + \omega \Delta X_i^t \tag{6}$$

where s, a and c represent the weights for separation, alignment and cohesion, respectively. F is the food factor, e is the enemy factor, ω is the inertia weight and t is the iteration vector. The location of the ith dragonfly at $(t + 1)$ is revised as [21]:

$$X_i^{t+1} = X_i^t + \Delta X_i^{t+1} \tag{7}$$

The radius of the neighbouring space increases as the algorithm progresses. If the dragonfly has no neighbours, then the position and velocity are updated using Levy flight technique, which is a variation of the random walk process to apply randomness to the position and velocity of dragonflies. The update can be performed as [5]:

$$X_i^{t+1} = X_i^t + \text{levy}(d) \times X_i^t \tag{8}$$

3.1 *Modelling of Objective Functions for Path Planning*

The main objective considered in this chapter is to navigate the autonomous vehicle in a road with different obstacles. The vehicle should sense the surroundings for obstacles and avoid them while reaching the destination with optimal smooth trajectory.

The problem of trajectory planning is considered as a minimization problem and two objective functions are considered in this chapter. The first function helps the vehicle to traverse to the destination avoiding the obstacles and second enables it to obtain a short smooth trajectory. The algorithm chooses the best path from the several paths for the vehicle to travel.

When the vehicle reaches the obstacle, the sensors detect it and the range of the obstacle is calculated. Dragonfly algorithm is then initiated to avoid the collision. It will find the best next position avoiding the obstacle while reaching the goal. The food source gives the optimal path for the ego vehicle to move. The position of the ego vehicle is updated as Eq. (6). The best approximations are stored and retrieved by an archive. The food source of the dragonfly is chosen from the archive.

Therefore, the food source is always a good candidate solution. Therefore, the quality of the food source is proportional to the optimal path length. The next position of the ego vehicle always depends on the distance between food source and goal and the obstacle. The important factors considered here are obstacle avoidance behaviour and goal-finding behaviour.

The collision avoidance is modelled using obstacle avoidance behaviour. The position of the food source is selected as the global best path, which satisfies the safety criteria of the system by maintaining the maximum distance between the ego vehicle and the obstacle. The objective function can be calculated as the Euclidean distance between the best position and the obstacle in the environment [23].

$$\text{Distance}_{obj-F} = \sqrt{(X_{ob} - X_{Fi})^2 + (Y_{ob} - Y_{Fi})^2} \quad (9)$$

where (X_{ob}, Y_{ob}) gives the position of the obstacle and (X_{Fi}, Y_{Fi}) gives the location of the food source.

The nearest obstacle to the vehicle is calculated as [23]:

$$\text{Distance}_{Obj-V} = \sqrt{(X_{ob} - X_V)^2 + (Y_{ob} - Y_V)^2} \quad (10)$$

The food source must be kept at the minimum distance from the goal. The food source gives the global best position. The goal-finding behaviour can be calculated as the Euclidean distance between the food source and the goal. It can be represented as [23]:

$$\text{Distance}_{G-F} = \sqrt{(X_G - X_{Fi})^2 + (Y_G - Y_{Fi})^2} \quad (11)$$

These two behaviours can be combined to find the objective function of the system. It can be represented as [23] :

$$\text{Objective function} = C1.1/\left(\min\left(\text{Distance}_{OB-F}\right)\right) + c2.\text{Distance}_{G-F} \quad (12)$$

where $C1$ and $C2$ are the fitting parameters, and they have a huge influence on the optimal trajectory.

3.2 Algorithm

The pseudo-code for the trajectory planning is given in Algorithm 1.

Algorithm 1 TP_Dragonfly

Input : Dragonfly Population P, Number of Dragonflies N
Output : Optimal Path Coordinates

```
Initialize the starting and goal positions of the autonomous vehicle.
Follow the reference trajectory until an obstacle is sensed
if vehicle senses obstacles, start dragonfly algorithm
Initialize the population of dragonflies randomly Pi (1,2,...N)
Set the step vectors ΔXi (1,2,..m)
while (t < Maximum number of iterations) do
    Calculate the fitness of each dragonfly f(Xi)
    Update the location of the food source and enemy
    Update the weights for inertia ω ,separation s, alignment a, cohesion c, food factor f and enemy factor e
    Compute S, A, C, F, and E using (1) – (5)
    Update neighbor radius, velocity vector, position vector using (6) – (8)
    Move the vehicle to the global best position
end while
```

4 Simulation Results

The proposed TP_Dragonfly algorithm for trajectory planning is simulated using Matlab. The lane-changing scenario is evaluated using the real data from the NGSIM (Next Generation Simulation) dataset. Next Generation Simulation (NGSIM) program is the project by US Federal Highway Administration and is collected by detailed vehicle trajectory data on southbound US 101 and Lankershim Boulevard in Los Angeles. The data from the NGSIM dataset provides real traffic information such as vehicle velocity, position, acceleration, lane, etc. This data is used for studying the features of lane-changing process and for validating the lane-changing models.

In the simulation, the positions, velocities and accelerations of ego vehicle and other vehicles including the obstacles are obtained from NGSIM dataset, and the optimal trajectory is generated by the proposed model. For each time step, the proposed method plans the trajectory of the ego vehicle dynamically. Figure 4 shows the trajectory planning using the proposed TP_Dragonfly algorithm.

This proposed method can be applied for both static and dynamic obstacles. When the ego vehicle senses a static obstacle, it starts changing the lane considering the safety criteria. But for the dynamic obstacles, when the speed of the ego vehicle is greater than the speed of the dynamic obstacle in front, then the ego vehicle starts the lane change. The ego vehicle follows the reference path until it senses the obstacle. When the vehicle detects the obstacle, then the TP_Dragonfly algorithm is

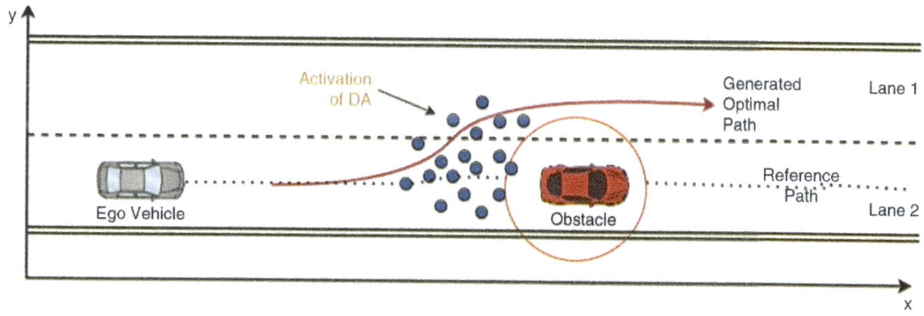

Fig. 4 Schematic diagram showing activation of TP_Dragonfly Algorithm

Table 1 Parameter values selected for TP_Dragonfly

Sl No	Symbol	Value
T	Max. Number of Iterations	430
N	Population Size	100
S	Separation	0.1
A	Alignment	0.1
C	Cohesion	0.7
F	Food	1
E	Enemy	1
C1	Fitting Parameter 1	1
C2	Fitting Parameter 2	1×10^{-4}
r1, r2	Random Numbers	[0,1]

activated to find an optimal path that avoids the obstacle. TP_Dragonfly algorithm calculates the next best position based on the objective functions and move forward avoiding the obstacle and reaches the destination.

The parameters used for simulation are shown in Table 1.

The efficiency of the proposed TP_Dragonfly algorithm relies on the accuracy of approximations of the parameters.

The proposed method can be validated using different lane-changing behaviours existing in the real world. The overtaking behaviour can be shown from the data given in the NGSIM dataset. This NGSIM data can be used to verify the efficiency of the proposed approach.

When the ego vehicle senses a static obstacle, it changes the lane to avoid collision. It initiates TP_Dragonfly algorithm to obtain the optimal path. Figure 5a and b shows the lane-changing behaviour of ego vehicle when a static obstacle is encountered.

When the speed of the ego vehicle is higher than that of the vehicle in front, the ego vehicle slowly changes the lane just before it approaches the vehicle in front. This vehicle is identified as a dynamic obstacle by the ego vehicle sensors, and it starts choosing a lane-change process by initiating TP_Dragonfly algorithm

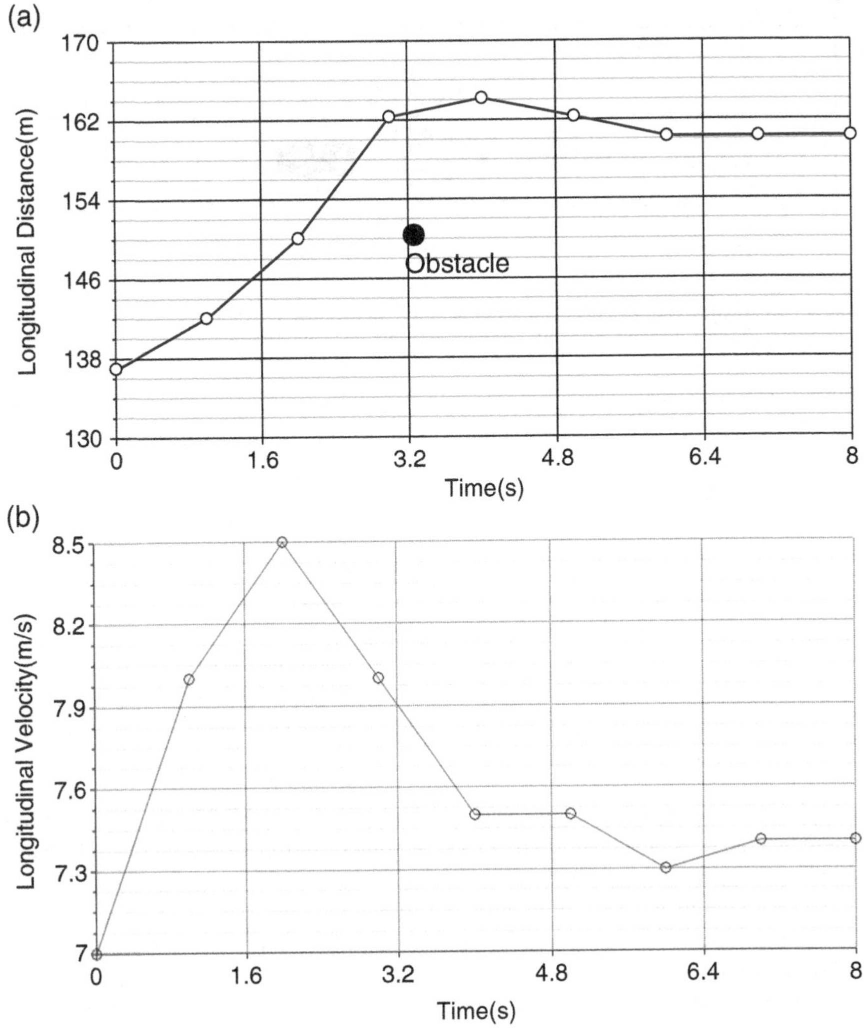

Fig. 5 Lane-change scenario for static obstacle

to generate an optimal path. Figure 6a shows the lane-changing behaviour of ego vehicle when it senses a dynamic obstacle.

Figure 6b shows the overtake scenario for the vehicle from the NGSIM dataset. The change of the speed of ego vehicle shows that the lane change adjusts the speed dynamically to adapt the change in the velocity of other vehicles.

The lateral positions and lateral velocity of the ego vehicle while changing the lane when sensing a dynamic obstacle are shown in Figs. 6c and d.

Figure 7 shows the lane-change trajectories generated by the proposed method and by the NGSIM data for the vehicle id 456 in NGSIM data. The vehicle detects a

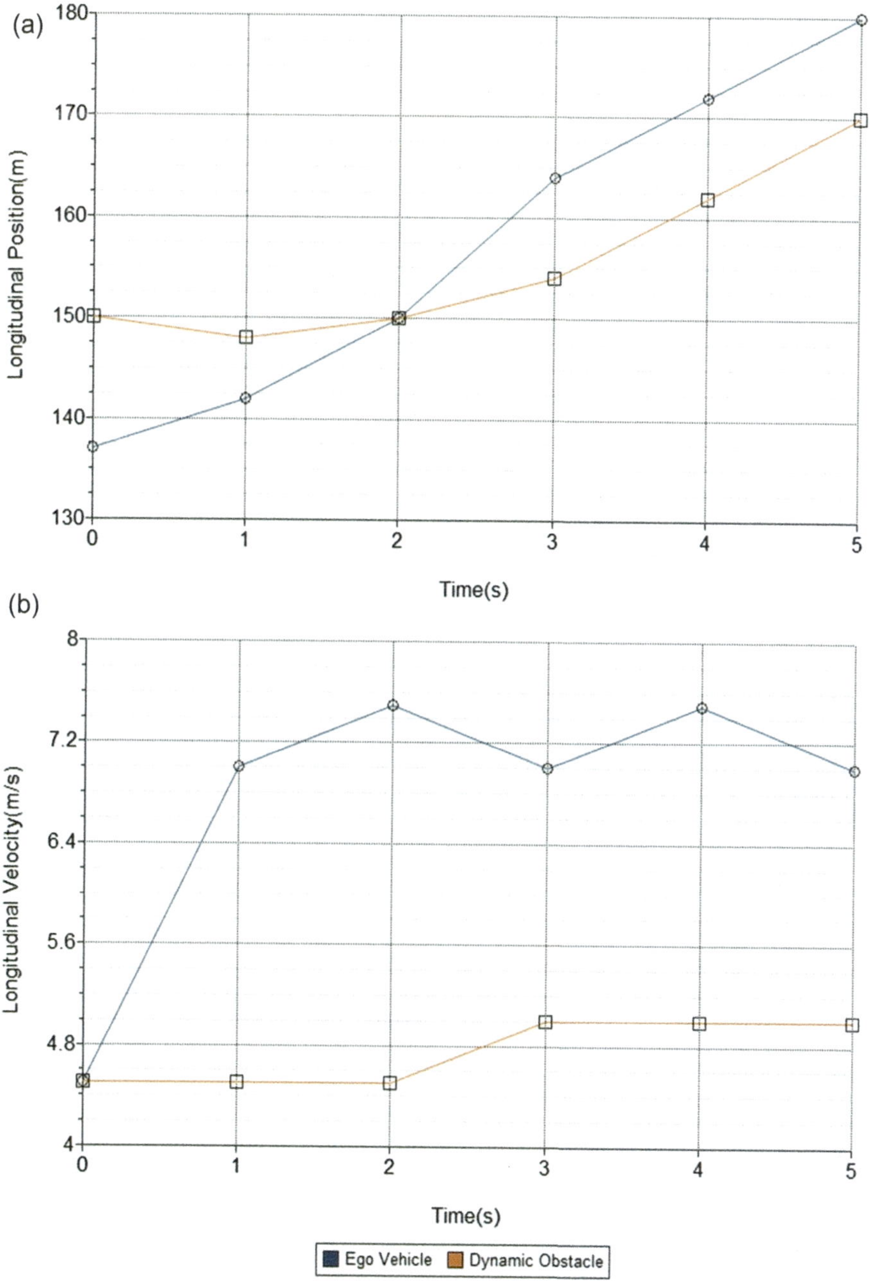

Fig. 6 (**a**) Lane-change scenario for dynamic obstacle. (**b**) Lane-changing scenario for dynamic obstacle. (**c**) Lane-change scenario for dynamic obstacle. (**d**) Lane-change scenario for dynamic obstacle

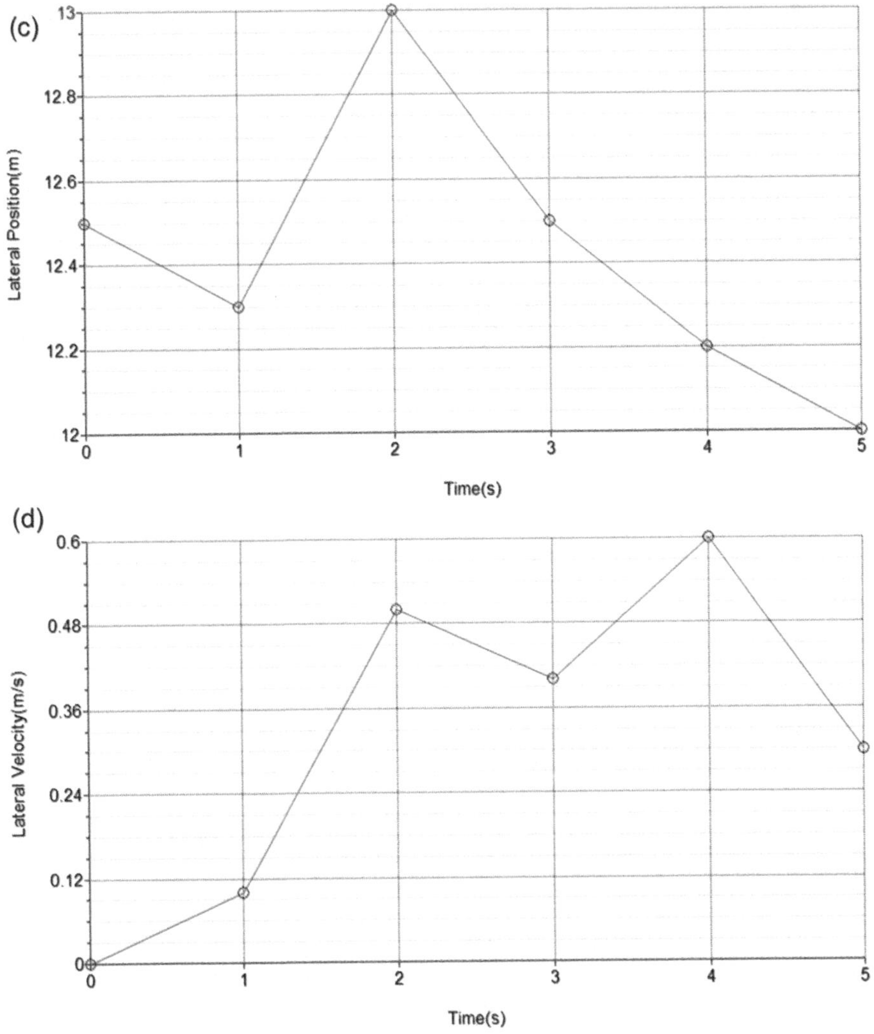

Fig. 6 (continued)

static obstacle, and it decides to change the lane. It is clear that the proposed method generates the optimal path.

The Fig. 8 compares the lane-change trajectories for vehicle taken from the NGSIM data for the vehicle id 466 in NGSIM data, which changes the lane as a result of the presence of moving obstacle. The ego vehicle moves at a speed of 60 km/h, and the obstacle vehicle moves at a speed of 45 km/h. The trajectory generated by the proposed method shows that it outperforms the real trajectory.

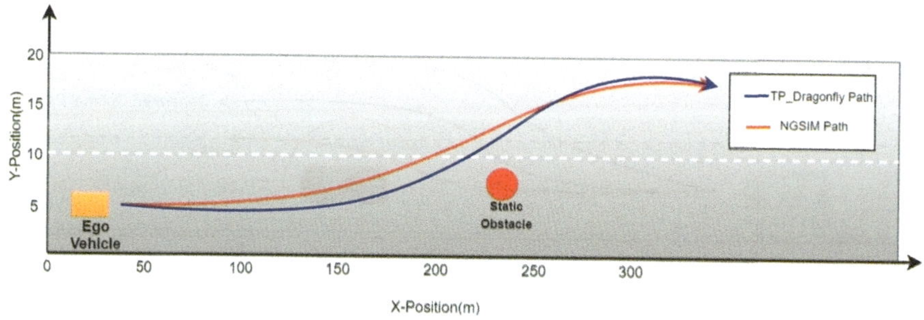

Fig. 7 Comparison of the trajectories generated for static obstacles

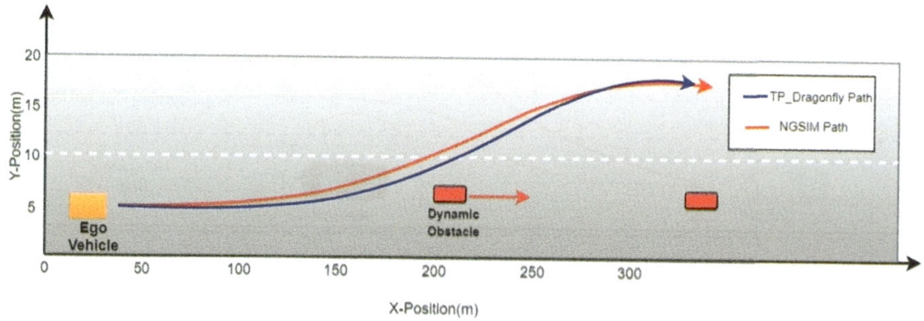

Fig. 8 Comparison of the trajectories generated for dynamic obstacles

The results show that the lane-change behaviour while sensing an obstacle in the current lane works well for the proposed TP_Dragonfly algorithm, which generates the optimal trajectory for lane change.

4.1 Influence of the Fitting Parameters on the Trajectory

Two fitting parameters C1 and C2 are used for the objective functions, and they have direct influence on the trajectory of the ego vehicle. The larger C1 value makes the vehicle move away from the obstacle and a smaller value makes it collide with the obstacle. The fitting parameter C2 decides the probability of the vehicle to reach the goal in optimal path. If C2 is large, there is high probability that the vehicle reaches the goal through an optimal path else it generates larger paths. The fitting parameters decide the convergence of the objective function, and the optimal values of these parameters eliminate the local minima. In this chapter, the fitting parameters are selected by trial and error. Figures 9 and 10 show the effect of C1 and C2 on the generated trajectory.

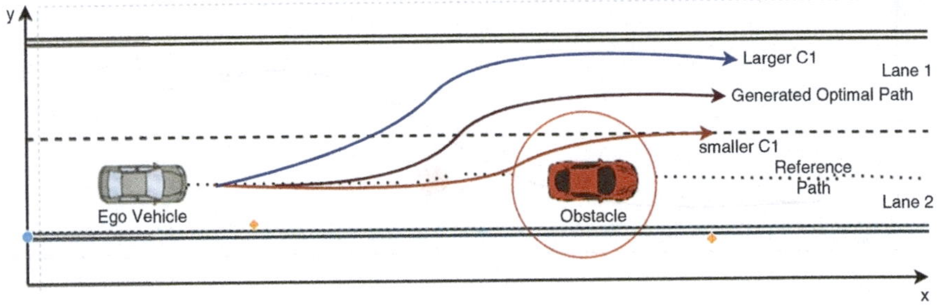

Fig. 9 Effect on fitting parameter C1 on trajectory

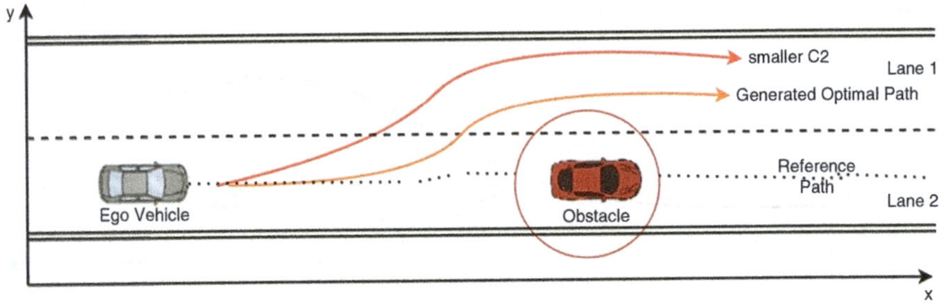

Fig. 10 Effect on fitting parameter C2 on trajectory

5 Conclusion

This chapter proposes an optimal trajectory planning technique for autonomous vehicles using a bio-inspired Dragonfly algorithm. The lane-change behaviour of the autonomous vehicles on approaching an obstacle is studied, and an optimal trajectory is generated for navigation to the goal for both static and dynamic obstacles. The simulation results show that this approach is feasible in the generation of optimal path for real-time data. The influence of fitting parameters on the trajectory is also investigated.

References

1. Meraihi, Y., Ramdane-Cherif, A., Acheli, D., & Mahseur, M. (2020). Dragonfly algorithm: a comprehensive review and applications. *Neural Computing and Applications, 32*, 16625.
2. Laghmara, H., et al. (2019). Obstacle Avoidance, Path Planning and Control for Autonomous Vehicles. In *IEEE Intelligent Vehicles Symposium (IV)* (pp. 529–534). Paris, France: IEEE.
3. Manipriya, S., Mala, C., & Mathew, S. (2020). *Significance of Real Time Systems in Intelligent Transportation Systems, Handling Priority Inversion in Time-Constrained Distributed Databases* (pp. 61–85). IGI Global Publications.

4. Manipriya, S., Mala, C., & Mathew, S. (2020). A Collaborative Framework for Traffic Information in Vehicular Adhoc Network Applications. *Journal of Internet Services and Information Security, 10*(3), 93.
5. Wang, Y., Lu, X., & Zuo, Z. (2019). Autonomous Vehicles Path Planning With Enhanced Ant Colony optimization. In *2019 Chinese Control Conference (CCC), Guangzhou* (pp. 6633–6638). Technical Committee on Control Theory, Chinese Association of Automation.
6. Nazarahari, M., Khanmirza, E., & Doostie, S. (2019). Multi-objective multi-robot path planning in continuous environment using an enhanced genetic algorithm. *Expert Systems with Applications, 115*, 106–120.
7. Mandava, R. K., Bondada, S., & Vundavilli, P. R. (2019). An Optimized Path Planning for the Mobile Robot Using Potential Field Method and PSO Algorithm. In *Advances in Intelligent Systems and Computing* (Vol. 817). Singapore: Springer.
8. Panda, M., Das, B., Subudhi, B., et al. (2020). A Comprehensive Review of Path Planning Algorithms for Autonomous Underwater Vehicles. *International Journal of Automation and Computing, 17*, 321–352.
9. Xue, J., Kawabata, K., Zhu, J., Ma, C., & Zheng, N. (2014). A Fast RRT Algorithm for Motion Planning of Autonomous Road Vehicles. In *17th IEEE International Conference on Intelligent Transportation Systems*. ITSC.
10. Liu, Q., Xu, H., Wang, L., Chen, J., Li, Y., & Xu, L. (2020). Application of Dijkstra Algorithm in Path Planning for Geomagnetic Navigation. In *IEEE 11th Sensor Array and Multichannel Signal Processing Workshop (SAM)* (pp. 1–4). Hangzhou, China: IEEE.
11. González, D., Pérez, J., Lattarulo, R., Milanés, V., & Nashashibi, F. (2014). Continuous curvature planning with obstacle avoidance capabilities in urban scenarios. In *17th International IEEE Conference on Intelligent Transportation Systems (ITSC)* (pp. 1430–1435). Qingdao: IEEE.
12. Daniel, J., Birouche, A., Lauffenburger, J.-P., & Basset, M. (2011). Navigation-based Constrained Trajectory Generation for Advanced Driver Assistance Systems. *International Journal of Vehicle Autonomous Systems (IJVAS), 9*, 269–296.
13. Hundelshausen, F., Himmelsbach, M., Hecker, F., Müller, A., & Wuensche, H.-J. (2009). Driving with Tentacles - Integral Structures for Sensing and Motion. In *The DARPA Urban Challenge* (pp. 393–440). Springer.
14. Mouhagir, H., Cherfaoui, V., Talj, R., Aioun, F., & Guillemard, F. (2017). Using evidential occupancy grid for vehicle trajectory planning under uncertainty with tentacles. In *IEEE 20th International Conference on Intelligent Transportation Systems (ITSC)* (pp. 1–7). Yokohama: IEEE.
15. Glaser, S., Vanholme, B., Mammar, S., Gruyer, D., & Nouvelière, L. (2020). Maneuver-based trajectory planning for highly autonomous vehicles on real road with traffic and driver interaction. *IEEE Transactions on Intelligent Transportation Systems, 11*(3), 589–606.
16. Zeng, D., et al. (2019). Novel Robust Lane Change Trajectory Planning Method for Autonomous Vehicle. In *IEEE Intelligent Vehicles Symposium (IV)* (pp. 486–493). Paris, France: IEEE.
17. Zhang, C., Chu, D., Liu, S., Deng, Z., Wu, C., & Su, X. (2019). Trajectory Planning and Tracking for Autonomous Vehicle Based on State Lattice and Model Predictive Control. *IEEE Intelligent Transportation Systems Magazine, 11*(2), 29–40.
18. Dolgov, D., Thrun, S., Montemerlo, M., & Diebel, J. (2008). Practical search techniques in path planning for autonomous driving. In *Proceedings of the First International Symposium on Search Techniques in Artificial Intelligence and Robotics (STAIR-08), (Chicago, USA)*. AAAI.
19. Zucker, M., Bagnell, J., Atkeson, C., & Kuffner, J. (2010). An optimization approach to rough terrain locomotion, Robotics and Automation (ICRA). In *IEEE International Conference* (pp. 3589–3595). IEEE.
20. Ben-Messaoud, W., Basset, M., Lauffenburger, J., & Orjuela, R. (2018). Smooth Obstacle Avoidance Path Planning for Autonomous Vehicles. In *IEEE International Conference on Vehicular Electronics and Safety (ICVES)* (pp. 1–6). Madrid: IEEE.

21. Mirjalili, S. (2016). Dragonfly algorithm: a new meta-heuristic optimization technique for solving single-objective, discrete, and multi-objective problems. *Neural Computing and Applications*,***27***, 1053–1073.
22. Mashadi, B., & Majidi, M. (2014). Global optimal path planning of an autonomous vehicle for overtaking a moving obstacle. *Latin American Journal of Solids and Structures, 11*, 2555.
23. Rahman, C. M., & Rashid, T. A. (2020). A survey on dragonfly algorithm and its applications in engineering. *ArXiv abs/2002.12126*.

A Novel Coherent Architecture for Traffic Signal Management in Internet of Things

S. Umaa Mageswari, C. Mala, and A. Santhana Vijayan

Abstract With recent developments in transportation, there is a rapid growth in vehicles leading to increase in traffic congestion. To manage this scenario, traffic management applications, such as Traffic Signal Management (TSM), traffic congestion, route guidance, speed advisory etc., based on Internet of Things (IoT) infrastructure are used. IoT is an overall system associating all the smart elements together and when confined with vehicles it is termed as Internet of Vehicles (IoV). The major challenge in IoV is the availability of information such as, traffic vehicle speed, density, volume headway etc. for these applications during low propagation delay or high network utility. To resolve these challenges, this chapter proposes a novel architecture namely Coherent Architecture for Traffic Signal Management in IoT (CATSMI) to manage traffic efficiently by improving and ensuring information availability. CATSMI uses recent advanced technology of Radio Frequency Identification (RFID), a non-intrusive sensor to support TSM developed using IoT. The performance of the proposed architecture is analysed using simulated values of IoV and RFID for TSM. The results show that the quality and precision of the Traffic Signal Management application with IoV have improved using the proposed CATSMI.

Keywords Internet of Things · Traffic signal management · Traffic information · Internet of vehicles · Coherent architecture · Radio frequency identification

---

S. Umaa Mageswari (✉) · C. Mala · A. Santhana Vijayan
Department of Computer Science and Engineering, National Institute of Technology, Tiruchirappalli, Tamil Nadu, India
e-mail: mala@nitt.edu; vijayana@nitt.edu

A. Haldorai et al. (eds.), *3rd EAI International Conference on Big Data Innovation for Sustainable Cognitive Computing*, EAI/Springer Innovations in Communication and Computing, https://doi.org/10.1007/978-3-030-78750-9_3

1 Introduction

Internet of Things (IoT) that uses vehicles as communicating node for information permits trade of data proficiently among vehicles and infrastructures which are described as IoV. Vehicles interact with one another furthermore, with public networks by means of V2V, V2I and V2P collaborations, which empowers both gathering of data and the real-time distribution of basic data about the status of the road structure [1, 2]. IoV innovation is more evolved than traditional intelligent transportation systems dependent on sensors and cameras all through road structure and control centre. The fast advancement of wireless intelligent innovations has supported the rise of IoV framework. The advancements associated with IoV are RFID, WSN, Zig Bee, Bluetooth, 6LoWPAN and different practices that have modelled M2M correspondence just as devoted correspondence innovations to vehicular condition, for example, 802.11p, WAVE and CALM. Intelligent uses of IoV can be commonly ordered into four kinds, i.e., traffic management, traffic safety, entertainment and service subscription [3–5]. The major challenges of IoV are security, privacy, trust management, communication latency, protocol verification, complexity of routing algorithm, environmental challenges, versatility, dependability, flexibility and dynamicity. The above-mentioned challenges may be reduced by using backup of necessary vehicle information-related RFID technology.

Radio Frequency Identification (RFID) is a technology whereby information or digital data is encoded in RFID tags, which are captured by a reader via radio waves. It is a minuscule electronic device that involves a little chip and a reception apparatus with an antenna. The improvements using RFID for traffic management application can be very well analysed using the popular and traffic signal update application. Traffic signals or traffic lights are situated at street convergences, person on foot intersections and different areas to control streams of traffic. The function of traffic signals is to give sophisticated coordination and control to guarantee that traffic moves as easily and securely as could reasonably be expected. Therefore, in this chapter, a pilot architecture for Traffic Signal Management using IoV with the proposed architecture of Coherent Architecture for Traffic Signal Management in IoT (CATSMI) is proposed. The architecture enables to provide an alternate source of traffic information from RFID technology as a backup to improve the reliability. After receiving the information from input resources, the control centre maintains data using DATEX to produce compatible output. The remainder of the chapter comprises the following sections. Section 2 discusses the related works of the proposed architecture. Section 3 presents the details of the proposed architecture. The simulation and performance analysis of the proposed architecture is presented in Sect. 4. Finally, the chapter is concluded in Sect. 5 followed by references.

2 Literature Survey

The Internet of Things (IoT) in traffic management and their existing challenges are discussed in this section. The alternate data resource for Radio Frequency Identification is efficiently used for traffic management applications. The existing Traffic Signal Management (TSM) using IoV is also discussed. In [6], IoT platform is used to designing and developing actual time needed for monitoring the traffic. Detection of vehicles in the intersection of lanes for various levels is done by using ultrasonic sensor, and the information getting from controller is sent to web server via Wi-Fi for analyse and storage purpose. Highest priority has been given to lane with heavy traffic. This information is given to priority system from the main system by using RF Transceivers. In [7], several identifications of behaviour of moving vehicles are described by a process model of the active identification using RFID. In this chapter, the zone is divided into segments based on inventory rounds. A matrix set is used for recording the details of unread, read and missed transponders in next rounds and segments. In [8], the authors proposed an efficient traffic management method by introducing an architecture and monitoring immediate response to traffic incidents.

In [5], the consumption time for sending a secured alert message to an emergency vehicle from other vehicles and also traffic signal control may be minimized by using optimized code and also by designing improved modulation and demodulation circuits for better data rates.

In [9], the authors proposed a method to overcome the traffic jam problem by assigning priorities to the ambulance and VIP vehicles during peak hours and give priority to heavy vehicles during night time.

In [10], the authors presented a vehicle identification method by using RFID and GSM compatible with IEEE802.15.4 for 2.4 GHz, for giving a full computerization of highway checking with distant observing station by utilizing two systems. The tag's CC2530F256 SoC MAC address is one of a kind used to produce a fixed holding up time, and the following system uses tag's CC2530F256 pseudorandom number generator by increasing the value of the main yield. It shields from the aggressors from getting to the label ID by assessing steady holding up season of the primary procedure.

In [11], the authors proposed low-latency and RFID calculation algorithm as elementary unit to help entangled issues for an enormous scope RFID organize. Express distinguishing proof is not required for every single label when utilizing this algorithm thus expel the latency congestion made by serialization during numerous entrance control. There might be dependable correspondence among tag and reader is expected. They introduced a series of calculations and their performance study under a disentangled memoryless loss channel model and stretched out them to deal with the effect due to backscattering impacts and connected losses created in functional RFID frameworks.

In [3], the authors optimized the duration of the traffic signal based on vehicle density for particular roadside by reducing waiting time for the drivers when crossing road signal. Clustering algorithms illustrations with KNN algorithm are used to determine vehicles count. In [4], the authors presented a five layered framework for IoV by examining the functionalities and their representations of all layers. An IoV network model is presented by using three numbers of network elements, such as cloud, communication network and client. In [12], the authors presented a bi-level improvement engineering; the high-level issue points the traffic light settings to limit the drivers' normal travel time, while the low-level issue targets accomplishing the system harmony utilizing the settings determined at the upper level. Genetic Algorithms with the combination of microscopic traffic simulation-based work are used to separate the complicate bi-level issue from manageable uni-level issue; this is illuminated successively.

In [13], the authors proposed IoT-based traffic congestion reducing system that automatically sets the signal operation time, which is decided by measuring the values of vehicle density at the road junctions. In [14], the authors proposed an IoT-based traffic management guidance, where traffic flow can be actively controlled by onsite traffic officials by using their smart mobile phones and also centrally monitored via Internet. In [15], the authors discussed various scenarios for improving this control by installing RFID readers and tags on road infrastructure and on vehicles, respectively, or vice versa. Authorized vehicles only allowed by verifying the information of the vehicles so controlling is possible to access zones with restrictions by using RFID. The above existing works discuss the TSM using IoV. It is seen that the existing work does not have the ability to overcome the lost information during failure of communication in IoV. To overcome the challenges in communication failure, there is a need of an integrated architecture to compensate the loss [1, 16, 17]. In order to compensate such drawback, this chapter proposes a pilot novel architecture known as Coherent Architecture for Traffic Signal Management in IoV (CATSMI) that uses the advantage of RFID as backup of information. The details of the proposed architecture are discussed in the subsequent sections.

3 Proposed Coherent Architecture for Traffic Signal Management in IoV (CATSMI)

To overcome the challenges of IoT described above, in this chapter, a Coherent Architecture for Traffic Signal Management in IoT is proposed. The proposed architecture is shown in Fig. 1. The real-time information from OBUs is communicated through V2I and V2V interfaces with RSU. Using multiple source of information for operability of any application enables to improve the precision of the outcome of the application. The most common current alternate source of traffic information is from video-based extractions. However, due to installation cost and maintenance cost of video-based inventory systems, the other sources for traffic information that can be effectively obtained are from Radio Frequency Identification (RFID) technology is

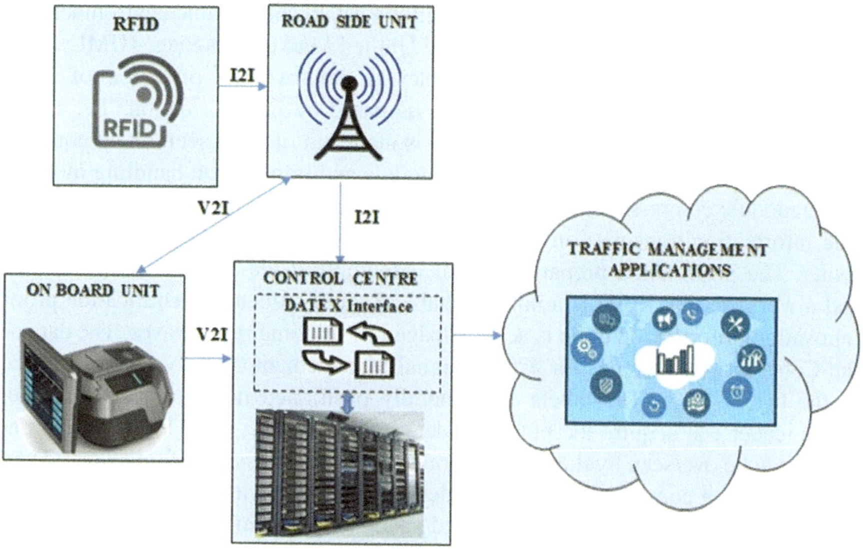

Fig. 1 Proposed architecture for CATSMI

utilized. To improve the information backup, this chapter proposes the extraction of traffic information using RFID. RFID is an experienced innovation that is effectively utilized in numerous enterprises to follow and oversee products and product. All the more explicitly, the most recent decade has seen a whirlwind of exploration for applying RFID innovation to the advancement of shrewd streets, fusing RFID to traffic the executive applications. RFID technology comprises tags that are acting as a tracking system that use smart barcodes in order to identify items. Essentially, it requires putting an RFID tag at each traffic sign or urban office, and vehicles, whose reason for existing is to examine RFID labels among its coverage area. The RFID is classified as active and passive based on the source of energy. RFID frequency ranges are classified into three types: (1) low frequency with 30–300 KHz, (2) high frequency with 3–30 MHZ and (3) ultra-high frequency 300 MHz–3 GHz.

RFID tags are very easy and simple to install and it requires storage capacity data up to 2 KB only. This technology cannot be easily replicated which drastically avoids security and attack-related issues. Using RFID technology to backup the traffic information aids to enhance the traffic managing application such as Traffic Signal Management that requires vehicle volume from the road segment. Upon receiving the traffic information from OBU and RFID through V2I communication interface, the control centre requires the data interface to have a generic format of data access. In the proposed architecture, the control centre maintains the data through DATEX interfacing which enable open exchange of traffic information across diverse traffic management applications [16–18]. DATEX is a standard developed to exchange data across different entities of transport system level for road segments. It allows the exchange of traffic information across multiple sources integration with operator co-operation. It allows establishing stable environment for system specification using the widely used Unified Markup Language (UML).

Finally, the proposed CATSMI enables to improve the precision of TSM application to the next level using the RFID-based traffic information. The traffic observing framework dependent on RFID is made out of three sections. It contains RFID module, information transmission module and information handling module. RFID module comprises a vehicle electronic tag, reader–writer and receiving wire. The information transmission module essentially comprises a coaxial cable and a router. The information preparing part is essentially made out of a control centre and a work station. RFID is a non-touchable Radio Frequency Identification proof innovation. Information trade is acknowledged by utilizing radio waves. The data of the Coherent Architecture for Traffic Signal Management objective item is stored in the tag or label. The labels are generally positioned in the objective article. RFID reader can acquire the put-away data by non-contact way. The label data in reader can be overseen by the work station framework and system framework. Then for information communication the reader can associate with the work station and arrange through a standard interface and sends the read information to the server through the work station to study and procedure.

4 Ontology for CATSMI

The ontology of the architecture enables to understand the details of the data, representations, relationships, roles and responsibilities of the entities. As in Fig. 2, the several traffic management applications are denoted as Application_id_1, Appli-

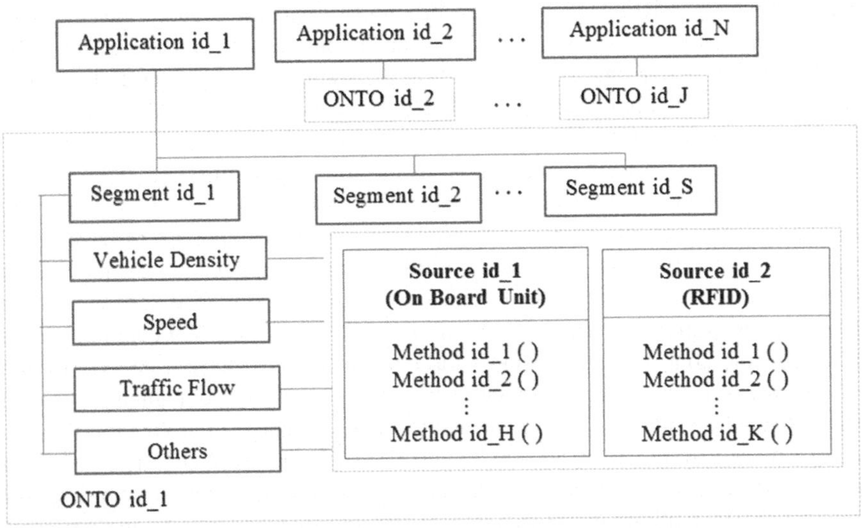

Fig. 2 Ontology for proposed CATSMI

cation_id_2, ... Application_id_N . The application requires traffic information for different segment of the roads for its operability. The segments of road are represented as Segment_id_1, Segment_id_2, ... Segment_id_S. For each of the segments, the traffic information such as volume, density and speed is obtained from two sources as per the proposed CATSMI. One source is the OBU from the smart vehicles of IoV. Another alternate source that is accessed through DATEX for the same traffic information is from RFID tags. The presence of alternate source enables in CATSMI aids to improve the quality and precision of applications that entirely depend on wireless communication of IoV that are prone to attacks and failure of information communication. Among several applications, the precision of Traffic Signal Management is further analysed to express the precision of the proposed work.

4.1 Traffic Signal Management

One of the popular real-time applications to manage traffic effectively is through traffic signal. To switch over the traffic between various directions across road segments, traffic signal phase cycles or timers are used for Traffic Signal Management (TSM). Most of the time, this may be fixed for a period of time leading to increase in waiting time of vehicle time inspite of low traffic density. For different traffic density, the phase cycle time needs to be changed automatically to minimize congestion and improve traveller experience. Using the proposed CATSMI, the traffic information to change the phase cycle time is obtained from OBU of IoV and RFID tags to ensure vehicle density may be accurate and consistent. Using CATSMI, traffic density pattern varied based on the result and also transit of vehicles speed may be improved during peak hours. As per the ontology in section 3, TSM comprises many OBU and RFID representing a separate road segment that communicates through a network to exchange information. The green band or the time for green phase of signal of each phase cycle is maintained consistent and updated based on the traffic density of the road segment. An algorithm based on [19] for estimating green band time is described below.

Algorithm for Estimating Green Phase Time for TSM

Step 1 The duration required to transfer or clear out vehicles in signal using start-up lost time and saturation headway as in [19] is determined using Eq. 1.

$$DT = L_1 + s.V \tag{1}$$

- where DT is the duration of time required to transfer or clear out V vehicles through signal,
- L_1 is the start-up lost time and
- s is the saturation headway in seconds.

Step 2 One of the road segments in the traffic signal is given default green phase which can be changed manually.

Step 3 The vehicle density obtained from the proposed CATSMI is categorized as low, medium, high and very high density, and their corresponding headway is calculated [2]. The maximum density using CATSMI is identified, and the time DT from Eq. 1 is calculated.

Step 4 The difference or variation in DT and previous green phase time is calculated.

Step 5 The current green phase time is calculated using the sum of variation in Step 4 and the previous green phase time.

Step 6 The green phase time is always maintained within a predefined minimum and maximum time limit.

Step 7 The green phase time for all the roads of the traffic signal is calculated in the descending order of traffic vehicle density using Steps 1–6 obtained from CATSMI.

5 Simulation and Performance Analysis

The performance of the proposed Coherent Architecture for Traffic Signal Management in IoV is analysed using the simulated values of traffic information. The traffic density required for Traffic Signal Management as in Sect. 3.2 from IoV and RFID is obtained from unique id obtained from PTV VISSIM. The screenshot of the simulation in VISSIM is shown in Fig. 3. Few observations of simulation of traffic

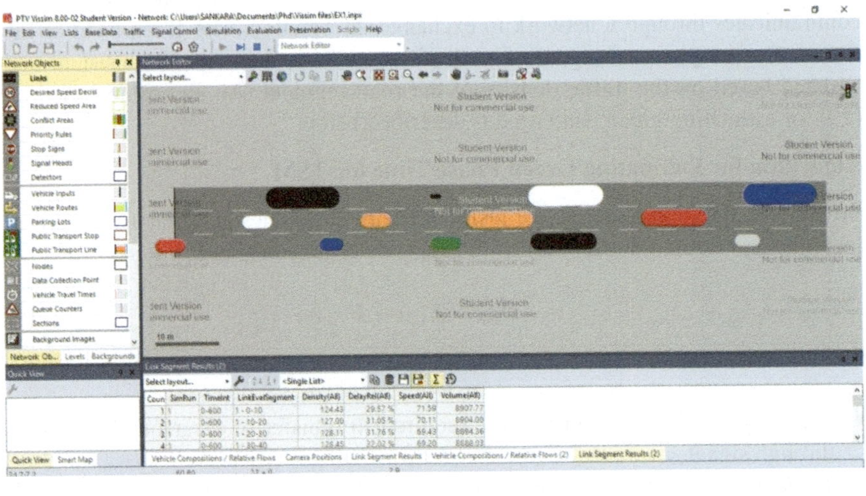

Fig. 3 Screenshot of simulation in VISSIM

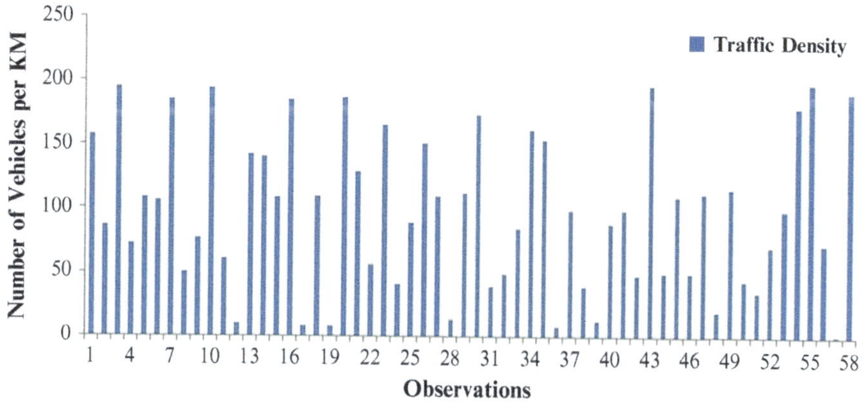

Fig. 4 Simulated values for traffic density

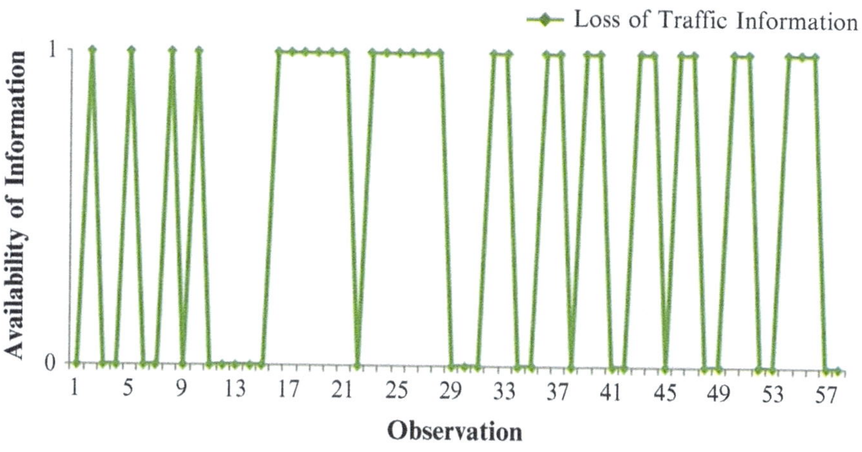

Fig. 5 Simulated values for loss of traffic information

density that are unanimously used for IoV and RFID are shown in Fig. 4. As the architecture requires to use the traffic density from alternate source of RFID, the simulated values of VISSIM for traffic density are assumed to be available without loss of information. The loss of information from IoV of OBU is simulated for several observations as in Fig. 5. The Eq. 1 is used to analyze the traffic information with loss of information, i.e., with CATSMI and without CATSMI. The result of the green phase cycle for low and high traffic density is shown in Fig. 6.

It is seen from the results that based on the loss of information from IoV in Fig. 5, the green band phase cycle becomes the default minimum value irrespective of the traffic density. However, using CATSMI, the green band phase compensates the lost data and shows a green phase time required to clear the vehicle represented as DT. Based on the simulated values, the precision of traffic signal green phase time is

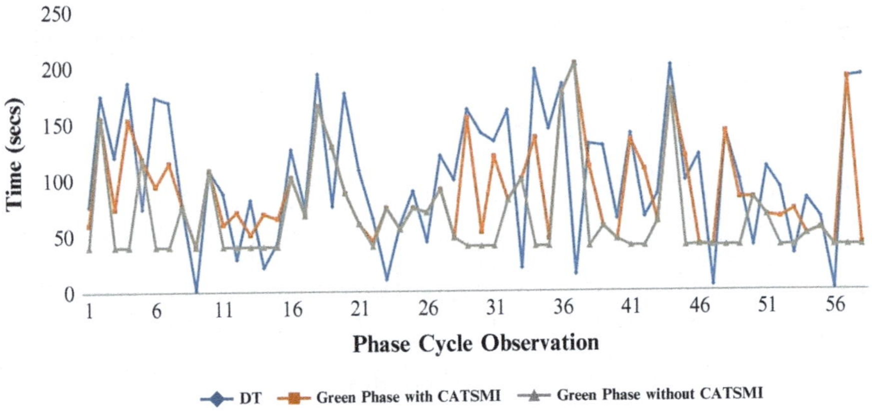

Fig. 6 Green phase for traffic signal using CATSMI

79.2% and 91.3% without and with CATSMI. Hence, the proposed architecture aids to improve the precision and quality of the application using traffic information from OBU of IoV.

6 Conclusion

Among several traffic management strategies using traffic information from On Board Unit (OBU) of Internet of Vehicles (IoV), Traffic Signal Management (TSM) is the most common application. In this chapter, a pilot novel Coherent Architecture for Traffic Signal Management using IoV (CATSMI) is proposed to improve the precision and quality of TSM. CATSMI uses the efficient Radio Frequency Identification (RFID) tags-based traffic information extraction to compensate the loss of information from IoV-based wireless communication. The performance of the proposed architecture is analysed using the simulated values of traffic density information from OBU and RFID for TSM. The results show that the green band phase cycle time for TSM shows improved and precise values with 91.3% accuracy using the proposed CATSMI.

References

1. Manipriya, S., Mala, C. & Mathew, S. (2019). Improved security schemes for efficient traffic management in vehicular ad-hoc network. In *The 4th International Symposium on Mobile Internet Security (MOBISEC), Communications in Computer Information Systems (CCIS) Proceedings*.

2. Manipriya, S., Mala, C. & Mathew, S. (2020). Significance of real time systems in intelligent transportation systems. In *Book chapter 5 in handling priority inversion in time-constrained distributed databases* (pp.61–85). IGI Global Publications, Scopus Indexed.
3. Janahan S. K., Veeramanickam, M. R. M., Arun, S., Narayanan, K., Anandan, R., & Javed, S. (2018). IoT based smart traffic signal monitoring system using vehicles counts. *International Journal of Engineering and Technology, 7*(221), 309.
4. Kaiwartya, O., et al. (2016). Internet of vehicles: Motivation, layered architecture, network model, challenges, and future aspects. *IEEE Access, 4*, 5356–5373.
5. Vijayaraman, P., & Jesu Jayarin, P (2019). Intelligent traffic management using RFID technology. *International Journal of Recent Technology and Engineering, 8*(4). ISSN: 2277-3878.
6. Nagmode, V. S., & Rajbhoj, S. M. (2017). An IoT platform for vehicle traffic monitoring system and controlling system based on priority. In *2017 International Conference on Computing, Communication, Control and Automation (ICCUBEA)* (pp. 1–5).
7. Pawłowicz, B., Trybus, B., Salach, M., Jankowski-Mihułowicz, P. (2020). Dynamic RFID identification in urban traffic management systems. *MDPI, Sensors, 20*, 4225.
8. Dandala, T. T., et al. (2017). Internet of vehicles for traffic management. In *IEEE International Conference on Computer, Communication and Signal Processing*. Chennai, India.
9. Chattaraj, A., Bansal, S., & Chandra, A. (2009). An intelligent traffic control system using RFID. *IEEE Potentials, 28*(3), 40–43.
10. Nafar, F., & Shamsi, H. (2018). Design and implementation of an RFID-GSM-based vehicle identification system on highways. *IEEE Sensors Journal, 18*(17), 7281-72-93.
11. Sze, W. K., Deng, Y., Lau, W. C., Kodialam, M., Nandagopal, T., & Yue, O. (2014). Channel-oblivious counting algorithms for large-scale RFID systems. *IEEE Transactions on Parallel and Distributed Systems, 26*(12):, 3303–3316.
12. Li, Z, Shahidehpour, M, Bahramirad, S., & Khodaei, A. (2017). Optimizing traffic signal settings in smart cities. *IEEE Transactions on Smart Grid, 8*(5), 2382–2393.
13. Sadhukhan, P., & Gazi, F. (2018, February). An IoT based intelligent traffic congestion control system for road crossings. In *2018 International Conference on Communication, Computing and Internet of Things (IC3IoT)* (pp. 371–375). IEEE.
14. Misbahuddin, S., Zubairi, J. A., Saggaf, A., Basuni, J., Sulaiman, A., & Al-Sofi, A. (2015). IoT based dynamic road traffic management for smart cities. *In 2015 12th International Conference on High-capacity Optical Networks and Enabling/Emerging Technologies (HONET)* (pp. 1–5). IEEE.
15. Pawłowicz, B., Salach, M., & Trybus, B. (2019). *Infrastructure of RFID-based smart city traffic control system: Progress in automation, robotics and measurement techniques* (Vol. 920). Cham: Springer.
16. Manipriya, S., Mala, C. & Mathew, S. (2020). Partition tolerant collaborative gateway architecture for interoperable ITS applications. *International Journal of Ad-hoc and Ubiquitous Computing, 35(4)*, 222–240.
17. Manipriya, S., Mala, C. & Mathew, S. (2020). A collaborative framework for traffic information in vehicular adhoc network applications. *Journal of Internet Services and Information Security, 10*(3), 93–109.
18. Datex: https://docs.datex2.eu/general/DATEXII-UserGuide_include.html
19. Tom, V., & Mathew, S. (2014). Design principles of traffic signal in transportation system engineering. In *NPTEL* (pp. 34.1–34.13).

Color-to-Grayscale Conversion for Images with Non-uniform Chromatic Distribution Using Multiple Regression

M. E. Paramasivam, R. S. Sabeenian, P. M. Dinesh, R. Anand, and Eldho Paul

Abstract Color-to-grayscale conversion methods try to identify weights for various color channels for obtaining a grayscale image. These weights can be either fixed globally or computed on a localized basis. This chapter presents an approach for computing the global weights using localized regions chosen using the assistance of Human Vision System. For a given image, the proposed method aims to maximize the required foreground information, which is normally present in the dominant color channel. The proposed method was tested on DIBCO 2013 dataset and qualitatively evaluated using PSNR, MSE, and SSIM. The results obtained have established to be more satisfactory. The experimental results of ours and other color-to-grayscale methods have been tabulated and discussed.

Keywords Color-to-grayscale conversion · Image analysis · Localized weight computation · Multiple regression · MSE · SSIM

1 Introduction

The advent of cost-viable and miniaturized semiconductors has enabled sensing and displaying of color images much easier in daily life. Despite such technological advancements, image analysis algorithms utilize color-to-grayscale (C2G) conversion. This step involves reducing a three-dimensional color image to a two-dimensional grayscale image. Care is taken during such a transformation, so that

This work is supported by Chockalingam Trust and Sona College of Technology.

M. E. Paramasivam (✉) · R. S. Sabeenian · P. M. Dinesh, · R. Anand · E. Paul
Department of Electronics and Communication Engineering, Sona College of Technology, Salem, Tamil Nadu, India
e-mail: sivam@sonatech.ac.in; sabeenian@sonatech.ac.in; dinesh@sonatech.ac.in; anand.r@sonatech.ac.in; eldhopaul@sonatech.ac.in
http://www.sonatech.ac.in/ece/paramasivam.html

A. Haldorai et al. (eds.), *3rd EAI International Conference on Big Data Innovation for Sustainable Cognitive Computing*, EAI/Springer Innovations in Communication and Computing, https://doi.org/10.1007/978-3-030-78750-9_4

every indispensable information in the color image is effectively mapped as a gray-level value. Few other constraints include perceptual mapping of alike color values with identical gray value along with maximum preservation of local luminance, color contrast, and Hue.

Loosely speaking, a color image can be considered as a distribution over three-dimensional integer space ($\mathbb{R}^{m \times n \times D}$). With m and n representing the number of rows and columns, respectively, D intends the third dimension space. In this chapter, we have considered the RGB [1, 2] color space model, despite the presence of many other [3] color representations. The RGB color model has the three primary colors, Red ($\mathcal{R}$), Green ($\mathcal{G}$), and Blue ($\mathcal{B}$) in $\mathbb{R}^{m \times n}$ space cascaded one on the other. Thus, the ultimate value of D is 3. The ultimate value for D becomes 3, ascertaining the primary chromatic components viz., Red ($\mathcal{R}$), Green ($\mathcal{G}$), and Blue ($\mathcal{B}$) colors.

A pixel in a color image can be expressed as $\mathcal{C}\{(i, j), k\}$, such that (i, j) represents the pixel coordinates in a $\mathbb{R}^{m \times n}$ space and k indicates the color channel ($k \in \{\mathcal{R}, \mathcal{G}, \mathcal{B}\}$) numerically denoted as ($k \in \{1, 2, 3\}$).

$$\mathcal{C}\{(i, j), k\} = \left[[\mathcal{R}(i, j) \quad \mathcal{G}(i, j) \quad \mathcal{B}(i, j)] : \mathcal{R}, \mathcal{G}, \mathcal{B} \in \mathbb{R}^{m \times n} \right], \forall (i, j)$$

such that $(1 < i < m)$ and $(1 < j < n)$.

C2G techniques are visualized as a dimension reduction problem ($\mathbb{R}^{m \times n \times 3} \rightarrow \mathbb{R}^{m \times n}$), such that every fine detail in the color image is effectively mapped to a unique grayscale value. A variety of solutions are available for carrying out dimension reduction in images. However, most of these methods drop off lively visual features and also effectively map the noise present in color image. The problem now gets supplemented as to increase the visual features of grayscale image.

Dimension reduction solutions for images generally happen with loss of vital information like contrast, brightness, etc. To solve this, a number of methods have been proposed, in which any one of the features is maximized to obtain a grayscale image. However, the color-to-grayscale (C2G) conversion process is a mandate in most of the image analysis algorithms. A circumstantial way of representing a color image is $\mathbb{C} \in \mathbb{R}^{m \times n \times D}$, where m and n represent the number of rows and columns, respectively, and D indicates the third dimension space. While there are a number of color mode representations [3], we have considered RGB [1, 2] color space model throughout this chapter. In such a case, the ultimate value for D is 3, such that the three planes are Red ($\mathcal{R}$), Green ($\mathcal{G}$), and Blue ($\mathcal{B}$) colors. A pixel of a color image ($\mathbb{C} = [\mathcal{R} \quad \mathcal{G} \quad \mathcal{B}]$) can be expressed as $\mathbb{C}\{(i, j), k\}$, such that (i, j) represents the pixel coordinates in a $\mathbb{R}^2$ space ($1 < i < m$ and $1 < j < n$), and k indicates the color channel ($k \in \{\mathcal{R}, \mathcal{G}, \mathcal{B}\}$).

To the best of our knowledge, we have not identified any method capable of carrying out a combined C2G and noise reduction. This chapter primarily tries to propose a perceptual computation of weights using Multiple Regression on regions as an integrated solution for C2G and noise removal. The method reported in this chapter tries to identify a global weight using localized features. Though the proposed method looks to be complex, it is computationally inexpensive.

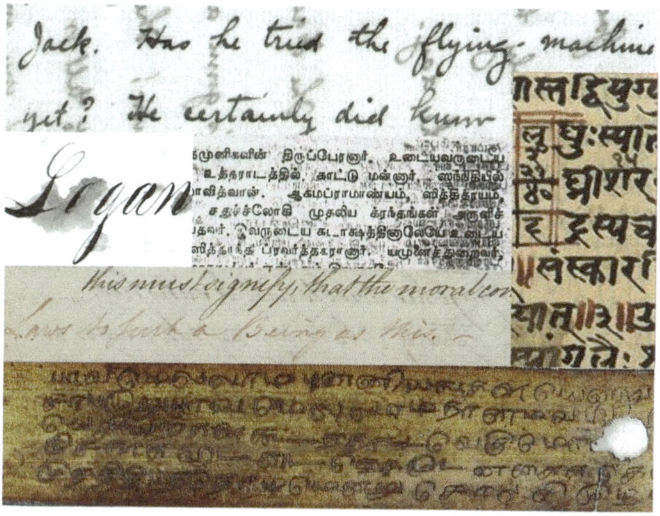

Fig. 1 A few samples of Historic Document Images with a variety of degradations

With an aim to assess effectiveness of the proposed noise removal C2G conversion, we have used Historic Document Image dataset—DIBCO [4–8]. The dataset has a total of sixteen images, with equal numbers in handwritten and printed category. The images are serially numbered separately with the handwritten ones prefixed with "HW" and the printed versions as "PR."

The motive behind choosing these datasets is due to the enormous presence of degradations [9] (bleed-through, inks/stains, character fading, etc.), which can be eliminated for effective segmentation of characters. Figure 1 shows a few sample images from DIBCO dataset, online repository of Marandog [10], NAMAMI [11], and a personal palm leaf manuscript collection of Marcel Chowriamah, MSK, Mauritius.

2 Previous Work

Chirstopher Kanan [12] has analyzed the effect of a dozen C2G methods for feature extraction on Face Recognition datasets. The methods listed in the above paper have ensured transformation of entire information present in the color image to a grayscale image. It is important to note that, this *entire information* also constitutes the noise present in the image. This section has concisely dealt with a few C2G techniques used for comparison with the proposed method.

Alexander Toet [13] classified the C2G algorithms into local and global mapping methods. With the former aiming at weighing variations of color on a localized basis, the later uses a uniform weight over the entire image. This paper has approached to compute the global weights using a few local weights of strategic regions for a given image.

2.1 Weighted Grayscale Conversion (WGC)

The most commonly used grayscale conversion is the **`rgb2gray`**[14] command of MATLAB, a Weighted C2G (WC2G) conversion, mathematically expressed as

$$\mathcal{L}_{WGC}(i,j) = [(0.3 \times \mathcal{R}(i,j) + 0.59 \times \mathcal{G}(i,j) + 0.11 \times \mathcal{B}(i,j))], \forall(i,j) \tag{1}$$

such that $\mathcal{R}, \mathcal{G}, \mathcal{B}, \mathcal{L}_{WGC} \in \mathbb{R}^{m \times n}$ and $\mathcal{R}, \mathcal{G}, \mathcal{B}$ represents the three primary color channels. Image editing software GIMP uses this method for gray-level transformation (Fig. 2).

2.2 Average

Pratt [15] has described the formation of grayscale image by computing the average of three chromatic components. A few texts have referred this method as *Luminance*.

$$\mathcal{L}_{Average}(i,j) = \left[\frac{1}{3}\left(\mathcal{R}(i,j) + \mathcal{G}(i,j) + \mathcal{B}(i,j)\right)\right], \forall(i,j) \tag{2}$$

The YUV and YIQ color models used for NTSC, PAL, and SECAM television transmissions utilize the above equation for computing the luminance component (Y) (Fig. 3).

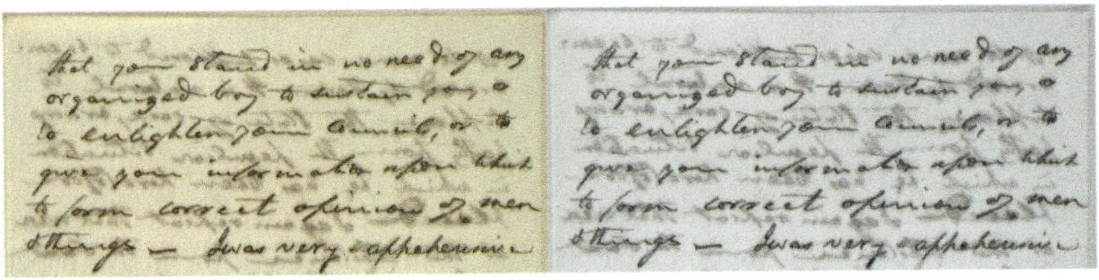

Fig. 2 Original color image and "rgb2gray" gray image for HW08—DIBCO 2013 dataset [SSIM:0.572, PSNR:12.911, MSE:0.051]

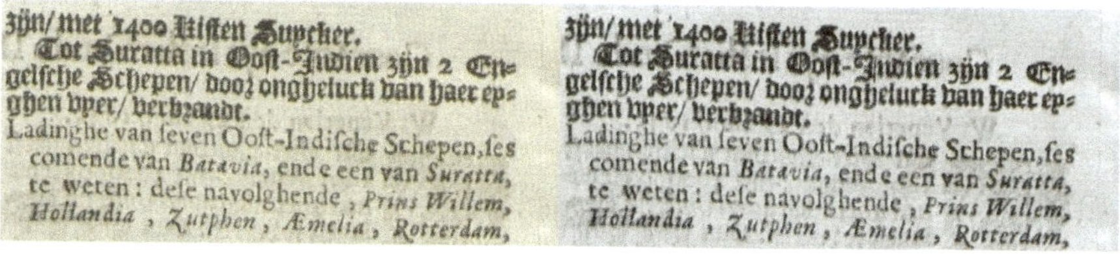

zijn/ met 1400 Kiſten Suycker.
Tot Suratta in Ooſt-Indien zijn 2 Engelſche Schepen/ door ongheluck van haer eyghen vyer/ verbrandt.
Ladinghe van ſeven Ooſt-Indiſche Schepen, ſes comende van *Batavia*, ende een van *Suratta*, te weten: deſe navolghende, *Prins Willem*, *Hollandia*, *Zutphen*, *Æmelia*, *Rotterdam*,

Fig. 3 Original color image and average gray image for PR07—DIBCO 2013 dataset [SSIM:0.583, PSNR:12.21, MSE:0.06]

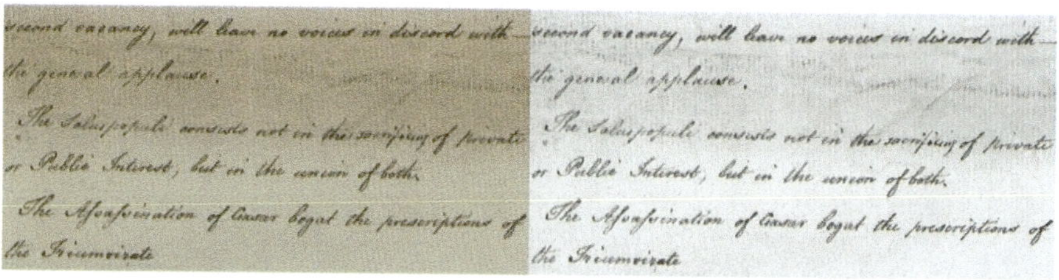

second vacancy, will have no voices in discord with the general applause.
The Salus populi consists not in the securing of private or Public Interest, but in the union of both.
The Assassination of Caesar begot the proscriptions of the Triumvirate

Fig. 4 Original color image and min–max gray image for HW02—DIBCO 2013 dataset [SSIM:0.682, PSNR:12.831, MSE:0.052]

2.3 Min–Max Grayscale Conversion

Another mathematical way of computing the grayscale value is to compute the average between most and least dominant among the three color channels for each pixel [12].

$$\mathcal{L}_{Min-Max}(i,j) = \left[\frac{\{\min(\mathcal{R}(i,j),\mathcal{G}(i,j),\mathcal{B}(i,j)) + \max(\mathcal{R}(i,j),\mathcal{G}(i,j),\mathcal{B}(i,j))\}}{2}\right], \forall(i,j) \tag{3}$$

Figure 4 shows the grayscale image formed by this method for HW02 named image in DIBCO 2013 dataset.

2.4 Optimized Solution for C2G

By using simple mathematical techniques, the above three methods have tried to maximize the visual perception features (contrast and brightness) in the perceived grayscale image. Qui et al. [16] have tried approaching the dimension reduction problem using optimization theory. The solution would be to maximize one of the first-order statistical features [Variance (σ^2)] of the perceived grayscale image.

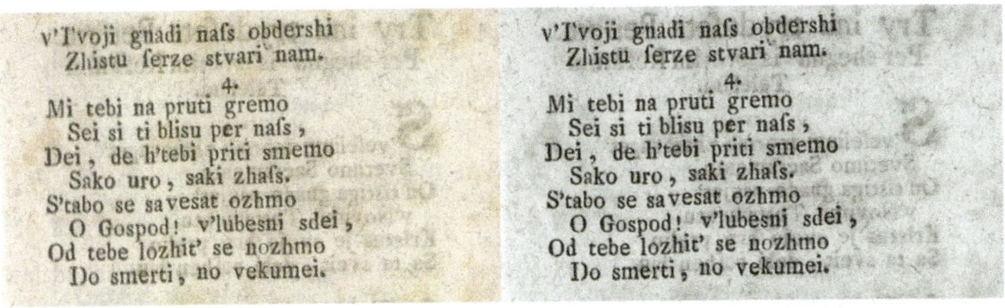
v'Tvoji gnadi nafs obdershi
Zhistu ferze stvari nam.
4.
Mi tebi na pruti gremo
Sei si ti blisu per nafs,
Dei, de h'tebi priti smemo
Sako uro, saki zhafs.
S'tabo se savesat ozhmo
O Gospod! v'lubesni sdei,
Od tebe lozhit' se nozhmo
Do smerti, no vekumei.

Fig. 5 Original color image and optimized gray image for PR02—DIBCO 2013 dataset [SSIM:0.543, PSNR:13.612, MSE:0.044]

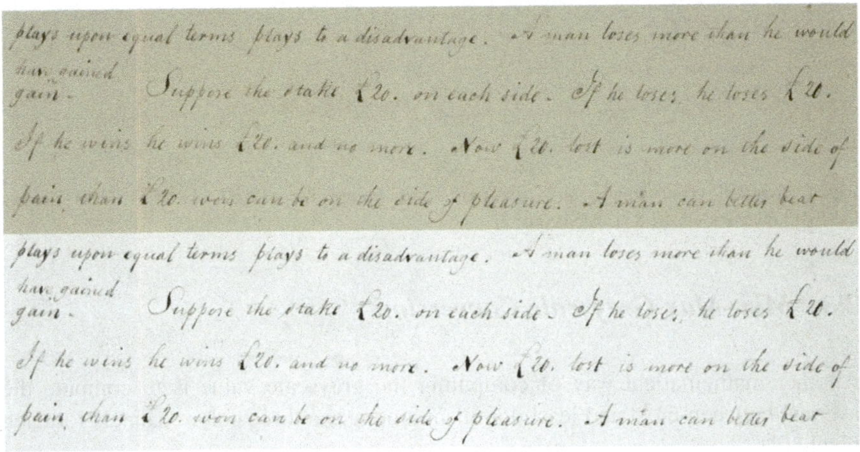
plays upon equal terms plays to a disadvantage. A man loses more than he would
have gained
gain. Suppose the stake £20. on each side. If he loses, he loses £20.
If he wins he wins £20. and no more. Now £20. lost is more on the side of
pain than £20. won can be on the side of pleasure. A man can better bear

Fig. 6 Original color image and YC_bC_r gray image for HW01—DIBCO 2013 dataset [SSIM:0.871, PSNR:16.29, MSE:0.023]

Micheal Ng [17] improvised the above model by adding up the contrast term to the problem definition. Figure 5 shows the optimized grayscale version for PR02 DIBCO 2013 dataset by maximizing variance and contrast.

2.5 Single Color Channel Visualization

For a given $\mathbb{R}^{m \times n \times 3}$ space, each color channel can be pictured in a $\mathbb{R}^{m \times n}$ space, thereby forming three grayscale images. One such visualization was done by Paramasivam and Sabeenian [18] for images in RGB color space. An alternate method of single color channel visualization is possible for images in YC_bC_r [19] color model. The Y channel for PR01 named image in DIBCO 2013 dataset is shown in Fig. 6 and can be expressed mathematically as

$$\mathcal{L}_{YC_bC_r}(i,j) = [16 + (65.739 \times \mathcal{R}(i,j) + 129.057 \times \mathcal{G}(i,j) + 25.064 \times \mathcal{B}(i,j))], \forall(i,j) \quad (4)$$

2.6 *Image Evaluation*

The classical methods of quantifying the noise present in an image are peak signal-to-noise ratio (PSNR) and mean squared error (MSE). In certain situations [20], these metrics are unable to extract the dominant features for exact quantification.

Structural Similarity Index Matrix (SSIM), a better qualitative and quantitative metric focusing on the structure, was established by Bovik [21]. Since our C2G conversion proposed in this chapter aims to extract the foreground structure, we have utilized SSIM for investigation, with reference to the ground truth available in DIBCO 2013 dataset.

The method proposed in this chapter shall make use of weighted color channel method. However, the weights shall not be fixed globally for all images but would be computed for every color image, such that the weights directly depend on the degree of color dominance in that particular image. The other parallel process is to understand the noise present in the image.

Modeling noise either mathematically or through any other means is still a great challenge. As an immediate solution, we have utilized the contribution of Human Vision System (HVS).

3 Proposed Method

This section shall elaborate on how Multiple Regression can be perceptually used to determine weights for each chromatic channel.

3.1 *Multiple Regression*

Regression is considered as a basic statistical approach of determining the relationship between a dependent variable Y and independent variable(s) X generally expressed as

$$Y \approxeq f(X, \beta) \quad (5)$$

where β in Eq. 5 represents the weighted coefficients, determined using a few samples of X.

If X is the only independent variable, the method is termed as Linear Regression, and Eq. 5 can be simplified as

$$Y \approxeq f(X, \beta) = \beta_1 + (\beta_2 \times X) \tag{6}$$

When X is a series set of values, i.e., $X = [X_1, X_2, X_3 \ldots X_t]$, such that each element in X is a column matrix of same size, then the function is expressed as

$$Y \approxeq f(X, \beta) = \beta_1 + (\beta_2 \times X_1) + (\beta_3 \times X_2) + \ldots + (\beta_t \times X_t) \tag{7}$$

Such a regression with more than one independent variable is termed as *Multiple Regression.*

The approach mentioned in this chapter has used each of the color channels ($\mathcal{R}$, $\mathcal{G}$, and $\mathcal{B}$) as an independent variable column matrix. Then, the grayscale image ($\mathcal{L}$) is expressed as

$$\mathcal{L} \approxeq f(\mathcal{C}, \xi) = (\omega_R \times \mathcal{R}) + (\omega_G \times \mathcal{G}) + (\omega_B \times \mathcal{B}) \tag{8}$$

The representation ξ in Eq. 8 denotes Dominant Chromatic Weight Matrix (DCWM). In the above equation, we have ignored the constant β_1 as seen in Eq. 7, due to the very minimal effect of this element in formation of grayscale image.

Expressing in the matrix form, we can rewrite Eq. 8 for all pixels in $\mathbb{R}^{m \times n \times 3}$ space as

$$\mathcal{L}(i, j) = [\xi \times \mathcal{C}(i, j)], \quad \forall (i, j) \tag{9}$$

where

$$\xi = \begin{bmatrix} \omega_R \\ \omega_G \\ \omega_B \end{bmatrix}, \mathcal{C}(i, j) = \begin{bmatrix} \mathcal{R}(i, j) & \mathcal{G}(i, j) & \mathcal{B}(i, j) \end{bmatrix}$$

To compute ξ matrix, the value for $\mathcal{L}$ can be taken up from any one of the methods discussed in Sect. 2. Equation 9 can be rearranged as shown in Eq. 10. In this chapter, we have used Weighted Grayscale Conversion described in Sect. 2.1 for computing ξ.

$$\xi = \begin{bmatrix} \omega_R \\ \omega_G \\ \omega_B \end{bmatrix} = \begin{bmatrix} \mathcal{R}(i, j) & \mathcal{G}(i, j) & \mathcal{B}(i, j) \end{bmatrix}^{-1} \times \mathcal{L}(i, j) = [\mathcal{C}(i, j)]^{-1} \times \mathcal{L}(i, j) \tag{10}$$

3.2 Dominant Chromatic Weight Matrix

This section shall detail on the various steps involved in identifying the Dominant Chromatic Matrix (ξ) from the computed Local Chromatic Weight Matrix (LCWM) Ψ. As a first step, we shall constitute the Ψ from various regions of a color image. Using the expertise of HVS, we shall identify five regions in the color image ($\upsilon 1$, $\upsilon 2$, $\upsilon 3$, $\upsilon 4$, and $\upsilon 5$). The approach of choosing various regions from the color image has been elaborated below. This is basically a guideline for choosing the regions, while the readers are allowed to choose other areas of the image conforming to the guideline.

3.2.1 Single Pixel Computation

For a given color image, exhibiting varied chromatic distributions, if a single pixel value is used for computing weights, a sparse matrix with weight exclusive for most dominant color is obtained. The highly dominant color channel is identified by computing the first-order statistical feature (Mean—μ) for each of $\mathbb{R}^{m\times n}$ space.

Consider the case of PR06 image in DIBCO 2013 dataset. The chromatic matrix at pixel position (658, 290) of PR08 image is found to be $\mathcal{C}(i,j) = [255 \quad 243 \quad 210]$. The corresponding grayscale value using Weighted Grayscale Conversion at the same pixel location is given as $\mathcal{L}(i,j) = 0.3 \times 255 + 0.59 \times 246 + 0.11 \times 210 = 242$.

This single pixel used for computation is considered as a region "$\upsilon 1$" in the proposed method. Using Eq. 10, we obtain the Local Chromatic Weight Matrix for this region as $\Psi_{\upsilon 1} = [0.9532 \quad 0 \quad 0]^T$.

In the above example, Red color channel ($\mathcal{R}$) exhibits the highest dominance, and hence, $\Psi_{\upsilon 1}$ matrix conforms to weight only for the first element. This computation would not be very realistic, and hence the weight matrix needs to be made fastidious. The contribution of other chromatic channels in the color image must be included in the formation of grayscale image at a relatively lower extent.

3.2.2 Column and Row Matrix Computation

To slightly improve the model, we shall now feed a row and column matrix containing the background region. We have limited the size to a maximum of 75, i.e., either the column matrix as (75×1) or the row matrix as (1×75). For PR06 image in DIBCO 2013 dataset, we have chosen the column matrix "$\upsilon 2$" from (396, 10) to (471, 10) and the row matrix "$\upsilon 3$" from (7, 289) to (7, 364). The weights for these regions are found to be $\Psi_{\upsilon 2} = [0.2540 \quad 0.6453 \quad 0.1053]^T$ and $\Psi_{\upsilon 3} = [0.2963 \quad 0.5967 \quad 0.1078]^T$.

3.2.3 Computation with Image Extracts

As a step to increase the dimensions of chosen regions, we have computed the weights for two handpicked $\mathbb{R}^{m\times n\times 3}$ spaces from PR06 image. The extracted chromatic channel pixels are modified into a column matrix and then computed for its corresponding weights. We have named these regions as $\upsilon 4$ and $\upsilon 5$. We suggest that the choice of these regions must be such that maximal foreground is acquired, with minimal background and noise as shown in Fig. 7. Computing weights for a 2D matrix is complex, and hence each color channel is converted to a column matrix for ease of calculation. The corresponding weights for these regions were found to be $\Psi_{\upsilon 4} = [0.3043 \quad 0.5828 \quad 0.1130]^T$ and $\Psi_{\upsilon 5} = [0.2983 \quad 0.5964 \quad 0.1050]^T$.

3.2.4 Local Chromatic Weight Matrix

The five regional weights obtained by the above steps are appended column-wise to attribute the Local Chromatic Weight Matrix [LCWM] (Ψ):

$$\Psi = \left[\Psi_{\upsilon 1}\ \Psi_{\upsilon 2}\ \Psi_{\upsilon 3}\ \Psi_{\upsilon 4}\ \Psi_{\upsilon 1}\right]$$

where

$$\Psi_{\upsilon 1} = \begin{bmatrix}\omega_R^{\upsilon 1}\\ \omega_G^{\upsilon 1}\\ \omega_B^{\upsilon 1}\end{bmatrix}\ \Psi_{\upsilon 2} = \begin{bmatrix}\omega_R^{\upsilon 2}\\ \omega_G^{\upsilon 2}\\ \omega_B^{\upsilon 2}\end{bmatrix}\ \Psi_{\upsilon 3} = \begin{bmatrix}\omega_R^{\upsilon 3}\\ \omega_G^{\upsilon 3}\\ \omega_B^{\upsilon 3}\end{bmatrix}\ \Psi_{\upsilon 4} = \begin{bmatrix}\omega_R^{\upsilon 4}\\ \omega_G^{\upsilon 4}\\ \omega_B^{\upsilon 4}\end{bmatrix}\ \Psi_{\upsilon 5} = \begin{bmatrix}\omega_R^{\upsilon 5}\\ \omega_G^{\upsilon 5}\\ \omega_B^{\upsilon 5}\end{bmatrix}$$

thus,

$$\Psi = \begin{bmatrix}0.9532 & 0.2540 & 0.2963 & 0.3043 & 0.2983\\ 0 & 0.6453 & 0.5967 & 0.5828 & 0.5964\\ 0 & 0.1053 & 0.1078 & 0.1130 & 0.1050\end{bmatrix} = \begin{bmatrix}\omega_R^{\upsilon}\\ \omega_G^{\upsilon}\\ \omega_B^{\upsilon}\end{bmatrix}$$

such that $\upsilon \in [\upsilon 1, \upsilon 2, \upsilon 3, \upsilon 4, \upsilon 5]$

Using the above Ψ matrix, the Dominant Chromatic Weight Matrix (ξ) can be obtained by identifying the maximum value in each row.

Fig. 7 $\upsilon 4$ and $\upsilon 5$ regions for image PR06 of DIBCO 2013

Table 1 Computation of DCWM from LCWM for PR06—DIBCO 2013 dataset

S. No.	Regions	Pixel coordinates	LCWM (Ψ)	DCWM (ξ)
1.	$\upsilon 1$	(658,260)	$[\mathbf{0.9532}\ \ 0\ \ 0]^T$	
2.	$\upsilon 2$	(396,10) to (471,10)	$[0.2540\ \ \mathbf{0.6453}\ \ 0.1053]^T$	$\begin{bmatrix} 0.9532 \\ 0.6453 \\ 0.1130 \end{bmatrix}$
3.	$\upsilon 3$	(7,289) to (7,364)	$[0.2963\ \ 0.5967\ \ 0.1078]^T$	
4.	$\upsilon 4$	(19,13) to (94,88)	$[0.3043\ \ 0.5828\ \ \mathbf{0.1130}]^T$	
5.	$\upsilon 5$	(13,99) to (88,174)	$[0.2983\ \ 0.5964\ \ 0.1050]^T$	

$$\xi = \begin{bmatrix} \max(\omega_R^{\upsilon}) \\ \max(\omega_G^{\upsilon}) \\ \max(\omega_B^{\upsilon}) \end{bmatrix} \tag{11}$$

Table 1 gives a clear picture of how the Dominant Chromatic Weight Matrix is obtained from Local Chromatic Weight Matrix. The Dominant Chromatic Matrix for the sample taken is constituted by identifying the maximum of among the LCWM (those indicated in bold).

4 Results and Discussions

Historic Document Images in $\mathbb{R}^{m \times n \times 3}$ space have a diversified contribution of the primary color channels $\{\mathcal{R}, \mathcal{G}, \mathcal{B}\}$. Paramasivam et al. [18] have demonstrated the availability of maximum Foreground Data (FD) in the most dominant color channel in Historic Document Images.

The C2G methods described in Sect. 2 have been computed on all sixteen images of DIBCO 2013 dataset for comparing with the grayscale image obtained by the proposed method. DIBCO contests were targeted to develop competitive algorithms that shall help in efficient binarization and thus support competent segmentation. Obtained grayscale images were validated for the availability of noise using ground truth provided by DIBCO contests.

The proposed method has provided appreciable results, such that the resultant images are almost close to the ground truth. Figure 8 shows the result of proposed method compared along with the classical Weighted Grayscale Conversion. To measure the structure similarity and the amount of noise eliminated in the proposed C2G method, we have computed SSIM, PSNR, and MSE. The value of SSIM generally lies between 0 and 1, with 0 indicating no similarity in the structure and 1 representing maximum relation to required structure. Table 2 contains the values of PSNR, MSE, and SSIM for Average, Min–Max, and Optimized C2G methods.

Table 3 contains the values of PSNR, MSE, and SSIM for Weighted Grayscale Conversion, YC_bC_r, and the proposed C2G methods. Since the method introduced in this chapter preserves the structure, values for SSIM are much higher when compared to the other five C2G methods.

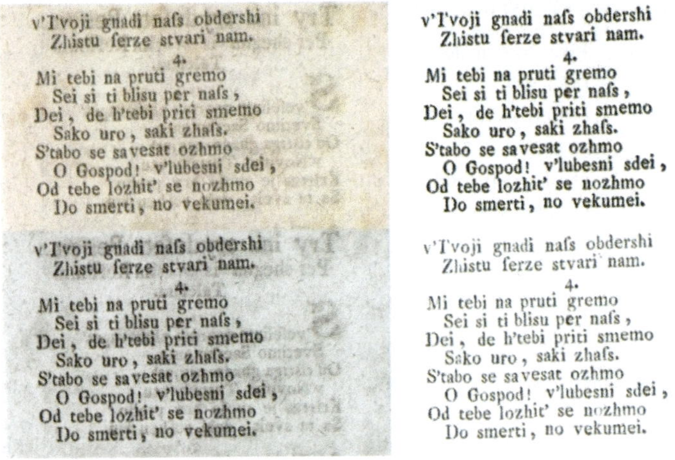

Fig. 8 PR02 of DIBCO 2013 dataset (from top left corner in anticlockwise: original image, ground truth, proposed method [SSIM = 0.872], weighted grayscale conversion [14] [SSIM = 0.546]

Table 2 Evaluation metrics of average [15], Min–Max [12], and optimized [17] C2G methods for handwritten and printed document images: DIBCO 2013 dataset

	Average			Min–Max			Optimized		
File name	PSNR	MSE	SSIM	PSNR	MSE	SSIM	PSNR	MSE	SSIM
HW01	16.267	0.024	0.872	16.302	0.023	0.870	16.274	0.024	0.873
HW02	13.133	0.049	0.694	12.831	0.052	0.682	12.942	0.051	0.689
HW03	15.050	0.031	0.758	14.953	0.032	0.755	15.074	0.031	0.764
HW04	13.600	0.044	0.453	13.600	0.044	0.453	13.600	0.044	0.453
HW05	15.403	0.029	0.470	15.403	0.029	0.470	15.403	0.029	0.470
HW06	17.168	0.019	0.762	17.168	0.019	0.762	17.168	0.019	0.762
HW07	13.961	0.040	0.851	13.961	0.040	0.851	13.961	0.040	0.851
HW08	12.400	0.058	0.569	12.250	0.060	0.554	12.553	0.056	0.570
PR01	12.787	0.053	0.539	12.853	0.052	0.542	13.609	0.044	0.578
PR02	13.612	0.044	0.543	13.662	0.043	0.539	13.678	0.043	0.543
PR03	12.282	0.059	0.738	12.474	0.057	0.737	12.457	0.057	0.738
PR04	9.534	0.111	0.636	9.444	0.114	0.629	9.554	0.111	0.635
PR05	11.942	0.064	0.649	11.790	0.066	0.643	11.986	0.063	0.650
PR06	12.828	0.052	0.597	12.355	0.058	0.586	12.765	0.053	0.603
PR07	12.210	0.060	0.583	12.145	0.061	0.579	12.238	0.060	0.585
PR08	10.619	0.087	0.461	10.322	0.093	0.456	11.028	0.079	0.467

Figure 9 demonstrates that the new method provides structurally better results when compared to other C2G methods, particularly for Historic Document Images containing enormous noise. To justify the allocation of weights for each of the color channels based on its dominance in $\mathbb{R}^{m \times n \times 3}$ space, we have tabulated (Table 4) the weights along with its corresponding mean of each color channel in $\mathbb{R}^{m \times n}$ space. It

Table 3 Evaluation Metrics of WGC [14], YC_bC_r [19], and Proposed Multiple Regression C2G methods for Handwritten and Printed Document Images: DIBCO 2013 Dataset

	WGC			YC_bC_r			Multiple regression		
File name	PSNR	MSE	SSIM	PSNR	MSE	SSIM	PSNR	MSE	SSIM
HW01	16.206	0.024	0.874	16.290	0.023	0.871	16.045	0.025	**0.912**
HW02	13.568	0.044	0.707	13.616	0.043	0.706	15.880	0.026	**0.873**
HW03	15.101	0.031	0.757	15.137	0.031	0.754	13.376	0.046	**0.852**
HW04	13.600	0.044	0.453	13.590	0.044	0.452	13.600	0.044	0.458
HW05	15.403	0.029	0.470	15.404	0.029	0.470	15.403	0.029	0.467
HW06	17.168	0.019	0.762	17.169	0.019	0.763	17.168	0.019	0.747
HW07	13.961	0.040	0.851	13.952	0.040	0.846	13.961	0.040	**0.800**
HW08	12.911	0.051	0.572	12.929	0.051	0.573	17.693	0.017	**0.908**
PR01	13.006	0.050	0.546	12.943	0.051	0.547	17.312	0.019	**0.899**
PR02	13.915	0.041	0.546	13.798	0.042	0.548	16.199	0.024	**0.872**
PR03	13.149	0.048	0.746	13.170	0.048	0.746	16.396	0.023	**0.885**
PR04	9.668	0.108	0.642	9.705	0.107	0.642	10.990	0.080	**0.765**
PR05	12.031	0.063	0.655	12.055	0.062	0.654	10.557	0.088	**0.670**
PR06	12.768	0.053	0.605	12.851	0.052	0.603	13.330	0.046	**0.754**
PR07	12.408	0.057	0.592	12.434	0.057	0.593	11.178	0.076	**0.687**
PR08	11.558	0.070	0.476	11.542	0.070	0.474	15.695	0.027	**0.846**

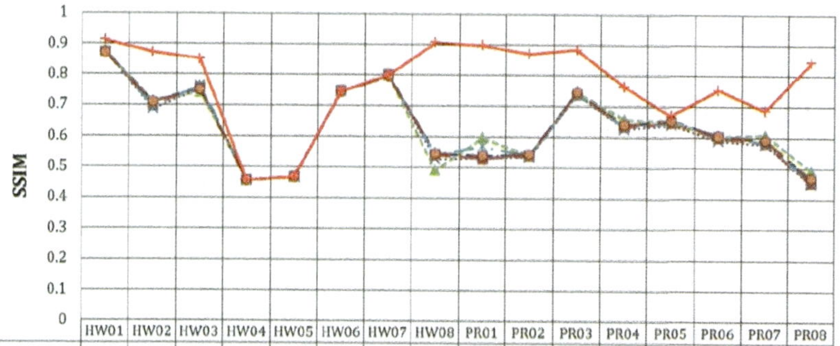

	HW01	HW02	HW03	HW04	HW05	HW06	HW07	HW08	PR01	PR02	PR03	PR04	PR05	PR06	PR07	PR08
rgb2gray	0.8744	0.7098	0.7562	0.4575	0.4675	0.7472	0.7999	0.5461	0.5374	0.5423	0.7452	0.6396	0.6542	0.6059	0.593	0.4666
Average	0.8727	0.6974	0.7557	0.4575	0.4675	0.7472	0.7999	0.5423	0.5314	0.5394	0.7379	0.6352	0.648	0.5997	0.583	0.4542
Max	0.874	0.7124	0.7436	0.4575	0.4675	0.7472	0.7999	0.4917	0.5977	0.5413	0.7441	0.6576	0.6525	0.6035	0.6089	0.4941
Min_Max	0.8711	0.6858	0.7536	0.4575	0.4675	0.7472	0.7999	0.5279	0.5341	0.5349	0.7376	0.628	0.6418	0.5891	0.5781	0.4495
Optimized	0.8731	0.6931	0.763	0.4575	0.4675	0.7472	0.7999	0.5429	0.5703	0.5401	0.738	0.6338	0.6489	0.6037	0.5846	0.4602
YCbCr	0.8702	0.709	0.7504	0.4565	0.4674	0.7468	0.7959	0.5457	0.5382	0.5438	0.7448	0.6394	0.653	0.6046	0.5926	0.4664
Mutiple Regression	0.9123	0.8729	0.8516	0.4575	0.4675	0.7472	0.7999	0.908	0.8991	0.8724	0.8855	0.7653	0.6703	0.7544	0.6869	0.8463

Fig. 9 SSIM comparison chart for various C2G conversion

is evident that the weights have been fixed based on the individual contribution of the color channels.

Consider the case of HW03; the computed weights are $[0.9660 \quad 0.5339 \quad 0.1889]^T$. Looking on the mean of each color channel, the highest is Red ($\mu_{\mathcal{R}} = 200.31$), and hence the correspondingly weight $\omega_{\mathcal{R}} = 0.9660$ also is the highest among the three weights. The other two values are noticeably varying with reference to the mean values ($\mu_{\mathcal{G}} = 191.38$ and $\mu_{\mathcal{B}} = 178.74$).

5 Conclusions

We have presented a color-to-grayscale conversion algorithm that uses Multiple Regression technique perceptually over different parts of the image. Such a method has proved to provide adaptive weights relatively depending on the contribution of each color channel. The results obtained by such methods are satisfactory both on a subjective (HVS) and on a quantitative (SSIM) basis. Table 4 clearly shows that the computed weights ω_R, ω_G, and ω_B are progressive with respect to the individual color channels, indicated by mean values.

5.1 Limitations of the Method

For images HW04, HW05, HW06, and HW07, the contribution of each color channel is equal. In such a case, the proposed method does not suit, and the weights have been contributed only for one color channel.

Table 4 Weights computed by the proposed method and mean for each color channel of images in DIBCO 2013 dataset

S. No.	File name	SSIM	ω_R	ω_G	ω_B	$\mu_{\mathcal{R}}$	$\mu_{\mathcal{G}}$	$\mu_{\mathcal{B}}$
1.	HW01	0.9123	0.9644	0.5287	0.1361	195.61	188.16	171.861
2.	HW02	0.8728	0.9414	0.6013	0.1492	177.02	163.95	140.50
3.	HW03	0.8516	0.9660	0.5339	0.1889	200.31	191.38	178.74
4.	HW04	0.4575	1	0	0	197.46	197.46	197.46
5.	HW05	0.4674	1	0	0	216.49	216.49	216.49
6.	HW06	0.7472	1	0	0	229.68	229.68	229.68
7.	HW07	0.7999	1	0	0	209.35	209.35	209.35
8.	HW08	0.9080	0.9481	0.5880	0.1164	205.88	192.45	159.55
9.	PR01	0.8990	0.9090	0.6018	0.1240	185.2652	167.19	143.43
10.	PR02	0.8724	0.94958	0.5911	0.1161	217.38	204.97	192.05
11.	PR03	0.8854	0.3037	0.9551	0.11604	197.10	216.24	190.90
12.	PR04	0.7653	1	0.5829	0.4545	193.10	179.89	162.07
13.	PR05	0.6702	0.9504	0.7646	0.1661	193.56	185.14	174.09
14.	PR06	0.7544	0.9532	0.6453	0.1130	200.04	188.21	162.52
15.	PR07	0.6868	0.9626	0.5678	0.1414	192.82	183.51	16,483
16.	PR08	0.8462	0.8713	0.5894	0.1161	209.21	17,762	119.11

References

1. Susstrunk, S., Buckley, R., & Swen, S. (1999). Standard RGB color spaces. In *In The Seventh Color Imaging Conference: Color Science, Systems, and Applications* (pp. 127–134).
2. Spaulding, K. E., Woolfe, G. J., & Giorgianni, E. J. (2000). Image states and standard color encodings (RIMM/ROMM RGB). In *Color and Imaging Conference (1)* (pp. 288–294). http://www.ingentaconnect.com/content/ist/cic/2000/00002000/00000001/art00052
3. Gonzalez, R. C., & Woods, R. E. (2006). *Digital image processing* (3rd ed.). Upper Saddle River, NJ: Prentice-Hall.
4. Gatos, B., Ntirogiannis, K., & Pratikakis, I. (2009). ICDAR 2009 Document Image Binarization Contest (DIBCO 2009). In *10th International Conference on Document Analysis and Recognition, 2009. ICDAR '09* (pp. 1375–1382). https://doi.org/10.1109/ICDAR.2009.246
5. Pratikakis, I., Gatos, B., & Ntirogiannis, K. (2020). H-DIBCO 2010—Handwritten Document Image Binarization Competition. In *Proceedings of the 2010 12th International Conference on Frontiers in Handwriting Recognition, ICFHR '10* (pp. 727–732). Washington, DC: IEEE Computer Society. https://doi.org/10.1109/ICFHR.2010.118
6. Pratikakis, I., Gatos, B., & Ntirogiannis, K.: ICDAR 2011 Document Image Binarization Contest (DIBCO 2011). In *2011 International Conference on Document Analysis and Recognition (ICDAR)* (pp. 1506–1510). https://doi.org/10.1109/ICDAR.2011.299
7. Pratikakis, I., Gatos, B., & Ntirogiannis, K. (2012). ICFHR 2012 Competition on Handwritten Document Image Binarization (H-DIBCO 2012). In *2012 International Conference on Frontiers in Handwriting Recognition (ICFHR)* (pp. 817–822). https://doi.org/10.1109/ICFHR.2012.216
8. Pratikakis, I., Gatos, B., & Ntirogiannis, K. (2013). ICDAR 2013 Document Image Binarization contest (DIBCO 2013). In *2013 12th International Conference on Document Analysis and Recognition (ICDAR)* (pp. 1471–1476). https://doi.org/10.1109/ICDAR.2013.219
9. Trier, O. D., & Taxt, T. (1995). Evaluation of binarization methods for document images. *IEEE Transactions on Pattern Analysis and Machine Intelligence , 17*, 312–315.
10. Maran Dog Online Repository of books, videos and audio files. Retrieved December 13, 2021 from http://www.marandog.info
11. National Mission for Manuscript. Retrieved December 13, 2021 from http://www.namami.org/
12. Kanan, C., & Cottrell, G. W. (2012). Color-to-grayscale: Does the method matter in image recognition? *PLoS ONE, 7*(1), e29740. https://doi.org/10.1371/journal.pone.0029740
13. Wu, T., Toet, A. (2014). Color-to-grayscale conversion through weighted multiresolution channel fusion. *Journal of Electronic Imaging, 23*(4). https://doi.org/10.1117/1.JEI.23.4.043004
14. Jayaraman, S., Esakkirajan, S., & Veerakumar, T. (2011). *Digital image processing*. New Delhi Tata Mc-Graw Hill Education Private Limited.
15. Pratt, W. K. (2001). *Digital image processing: PIKS inside* (3rd ed.). New York, NY: Wiley.
16. Qiu, M., Finlayson, G. D., & Qiu, G. (2008). Contrast maximizing and brightness preserving color to grayscale image conversion. In *Conference on Colour in Graphics, Imaging, and Vision (1)* (pp. 347–351). http://www.ingentaconnect.com/content/ist/cgiv/2008/00002008/00000001/art00075
17. Zhengmeng, J., & Michael Ng, K. (2015). A contrast maximization method for color-to-grayscale conversion. *Multidimensional Systems and Signal Processing, 26*(3), 869–877. https://doi.org/10.1007/s11045-014-0295-2
18. Paramasivam, M. E., & Sabeenian, R. S. (2015). Contrast based color plane selection for binarization of historical document images. In *2nd International Conference on Emerging Trends in Electrical, Communication and Information Technologies (ICECIT 2015)*.
19. Nixon, M., & Aguado, A.S. (2008). *Feature extraction & image processing* (2nd ed.). Academic Press.

20. Wang, Z., & Bovik, A. C. (2009). Mean squared error: Love it or leave it? A new look at signal fidelity measures. *IEEE Signal Processing Magazine, 26*(1), 98–117. https://doi.org/10.1109/MSP.2008.930649
21. Wang, Z., Bovik, A. C., Sheikh, H. R., & Simoncelli, E. P. (2004). Image quality assessment: From error visibility to structural similarity. *IEEE Transactions on Image Processing, 13*(4), 600–612. https://doi.org/10.1109/TIP.2003.819861

Concurrent Spatial Color Information Processing for Video-Based Vehicle Detection Applications

S. Manipriya , C. Mala, and Samson Mathew

Abstract The increase in the number of vehicles requires an efficient and well-organized solution for traffic management and regulation. Among several traffic management systems of Intelligent Transportation Systems (ITS), video-based system has gained wide popularity that uses computer vision techniques. One of the significant processes of computer vision technology is to detect vehicle from traffic videos. Effective image processing-based computer vision techniques are available to detect vehicles. The accuracy of the detection primarily depends on the section the video being processed for detection, which is completely neglected by the existing image processing technique. Hence, in this chapter, a Concurrent Spatial Color Information Processing System (CSCIPS) is proposed that uses a novel detection framework known as Virtual Multi-Loop Crates (VMLC). The proposed framework VMLC uses all the spatial color information for image processing technique concurrently without loss of any information. The performance of the proposed CSCIPS system with VMLC is analyzed with background subtraction-based image processing techniques in benchmark traffic videos. The results show better performance in processing time, storage space, and accuracy for vehicle detection when compared to other existing frameworks.

Keywords Traffic management · Intelligent transportation systems · Vehicle detection · Virtual multi-loop crates · Image processing · Spatial color information · Background subtraction

---

S. Manipriya (✉) · C. Mala
Department of Computer Science and Engineering, National Institute of Technology, Tiruchirappalli, Tamil Nadu, India
e-mail: mala@nitt.edu

S. Mathew
Department of Civil Engineering, National Institute of Technology, Tiruchirappalli, Tamil Nadu, India
e-mail: sams@nitt.edu

A. Haldorai et al. (eds.), *3rd EAI International Conference on Big Data Innovation for Sustainable Cognitive Computing*, EAI/Springer Innovations in Communication and Computing, https://doi.org/10.1007/978-3-030-78750-9_5

1 Introduction

Intelligent Transport Systems (ITS) is the prominent and recent technologically advanced research field for regulating traffic efficiently. At present, the number of vehicles has been exponentially increased with constant development in urban roads, which intensifies the importance of managing it efficiently. Video cameras have been used to monitor traffic because of their rich contextual information for visualization and easy perception of human beings. Computer vision- or video/image-based analysis offers a faster and more accurate solution to many transportation systems. For any video-based ITS application, vehicle detection becomes one of the major significant tasks to enumerate traffic statistics such as traffic volume, headway, speed, density, occupancy, classification, etc. [1–3]. Traffic video-based applications are classified as traffic flow-, safety-, and enforcement-based applications. Some of the popular applications include congestion analysis, occupancy analysis, traffic count, incident detection and change warning, pedestrian detection, speed enforcements, intersection enforcements, desired vehicle detection, etc. [4]. These applications provide an effective way to manage, monitor, and regulate transport system in order to make coordinated, safer, and efficient use of the transportation system. For traffic status enumeration, the effectiveness of the applications depends largely on the appropriateness of vehicle detection [17].

The major factors that affect the precision of vehicle detection [1–3, 5, 6] are (1) dependency on camera properties or the video content, (2) characteristics of the video environment, i.e., illumination and weather changes, (3) time complexity involved in processing a video, and (4) efficiency and robustness of the image processing techniques. Apart from these factors, the major criterion for vehicle detection that is ignored in literature is the need and process of eliminating the unwanted content of the video that contributes to the false positives in detection [7, 8]. Therefore, a detection method must be established that is reliable enough to identify the vehicle under any conditions with minimal false detections. This chapter focuses on easily integrable novel framework for any video-based vehicle detection application for ITS. The novel framework known as Virtual Multi-Loop Crates (VMLC) is proposed that uses Concurrent Spatial Color Information Processing System (CSCIPS) for improving the precision and robustness of vehicle detection.

The rest of the chapter is arranged as follows: the existing framework and detection processes are discussed in Sect. 2. Section 3 presents the descriptions of the proposed detection framework, its significant features, and novel Concurrent Spatial Color Information Processing System (CSCIPS). Section 4 addresses evaluation outcomes and VMLC performance analysis with other frameworks and detection methods. Finally, the conclusion is given in Sect. 5 followed by Acknowledgement and References.

2 Literature Survey

Over more than two decades, several methods for detecting vehicle from videos have been introduced to assess traffic parameters. Since the methods proposed in this area are a broad area of research, few popular methods are discussed. Most methods use one or two extracted features such as color, texture, shape, corners, edge, shadows, or high-level features such as Haar-like features, Scale Invariant Feature Transform (SIFT), etc. to define and classify vehicles using appearance or motion of the vehicle. The authors of [1, 2, 9–11] presented a detailed report on existing vehicle detection approaches with their application in ITS. It presents the appearance and motion-based detection methods under varying environmental conditions.

In appearance-based methods, the moving objects are identified and detected by training the system with prior knowledge of the object properties and characteristics by using machine or deep learning techniques. Deep learning and machine learning methods such as Adaptive neuro-fuzzy inference system, Support Vector Machine (SVM), Neural Network, Ada Boost, and other appearance-based methods are available to detect possible vehicle candidates using the abovementioned features [2, 9, 11, 12]. There are other state-of the-art technologies in detecting moving object presented in detail in [4]. Few of them include The Deformable Parts Model (DPM), Convolution Neural Network (CNN) in [9], Region with features of CNN (R-CNN), Fast R-CNN, Region Proposal Network (RPN), You Only Look Once (YOLO), Evolving Boxes (EB).

Other approaches in motion-based techniques use pixel intensities and its properties to distinct background and moving content of the video, and the process is known as Background Subtraction (BS). BS techniques are the most common and popular for motion-based vehicle detection. It works on the concept of identifying the moving object by subtracting the background image and foreground or current image. The background image has to reflect the actual background of the traffic scene apart from moving vehicles. Hence, the major precision of detecting object in BS depends on the modeling of background image [5, 13, 14]. Also, the background image has to adapt to the dynamic changes over time in real life such as camera oscillations, trees, weather changes, etc. BS algorithms are classified as parametric and non-parametric methods for background image modeling. Detailed survey and reviews of the BS algorithms based on the accuracy, storage space, and complexity are discussed in [6, 15, 16]. Few of the most popular methods for BS include running average, Mixture of Gaussian (MOG), VIBE, Codebook, Kernel Density Method, etc. [1, 5, 17].

To sum up the overall existing works, the current systems in vehicle detection have various image processing methods and approaches that are specific to an application or properties of traffic video such as camera quality, angle, time of video recordings, etc. Consequently, it is essential to identify the suitable techniques that overcome the challenges of having a successful operational system with optimal balance in robustness, accuracy, and computational cost. To improve the robustness and efficiency of the detection process, it is necessary to eliminate the complexity of the video as much as possible. Almost all the existing methods use the entire image for processing which adds up the complexity of the detection process and only very few works use frameworks such as grids and selected region [1, 7, 8]. In reality, the detection of vehicles for traffic information is an integral and inevitable process of ITS applications that require solution to be simple, adaptive, and precise detection. To handle the aforementioned short comings, this chapter proposes a novel framework or structure with Virtual Multi-Loop Crates (VMLC) that uses Concurrent Spatial Color Information Processing System (CSCIPS).

3 Proposed Vehicle Detection Process with Virtual Multi-Loop Crates (VMLC) Using Concurrent Spatial Color Information Processing System (CSCIPS)

Existing vehicle detection algorithms rarely consider the environment of videos captured. The aim of installing cameras is not only to collect traffic data but is also most commonly used for surveillance, which obviously contributes to different backgrounds as shown in Fig. 1. Subsequently, due to the nature of the video, there is a need for a framework that removes the real world complexities.

Fig. 1 Diverse traffic videos from surveillance cameras

3.1 *Virtual Multi-Loop Crates (VMLC)*

A novel vehicle detection framework known as Virtual Multi-Loop Crates (VMLC) is proposed to eliminate the unwanted disturbances and complications in traffic video background. As first step to the proposed vehicle detection process, the first frame of the video is provided to the user to select the Significant Region of Frame (SRF). The user needs to use click and drag functionality of the mouse to do the same. The selected region is converted as an adjacent connected virtual segments or crates as shown in Fig. 2. It comprises several layers of segments with grid-like patterns that are continuous and connected. Individual segments are denoted as crates and a sequence of crates forms a loop. The direction of movement is automatically identified by using the sequence of variation in histogram of crates of virtual loops. Variation of histogram values in the adjacent crates of the multi-loop aids in categorizing the movement easily. The set of middle crates perpendicular to the direction of motion forms the Middle Layer Loop (ML2). All the detection process occurs only in ML2 as in Fig. 2. The main advantage of multi-loop is to detect movement in any direction of a surveillance video. The proposed framework is attempted to handle traffic surveillance-based videos to detect vehicle apart from the dedicated detection cameras. Most of these videos have one or more directions of vehicle movement. When SRF is defined for such videos, it eliminates several complexities such as reducing the entire frame for computation, restrict the detection process to a specific direction of motion, etc. However, the other detection process does not have the option to choose the detection technique specific to the direction of vehicle movement. Several existing detection methods scan the entire image or detect in specific region or sliding window or spatial pooling as the videos are accessed with prior knowledge of the camera properties. Using such algorithm in a multi-directional movement of vehicles such as surveillance video or unforeseen traffic content mostly results in reduced detection performance by detecting unwanted or all moving objects. Regardless of detection accuracy, such detection becomes pointless to any ITS application [5]. As most ITS applications

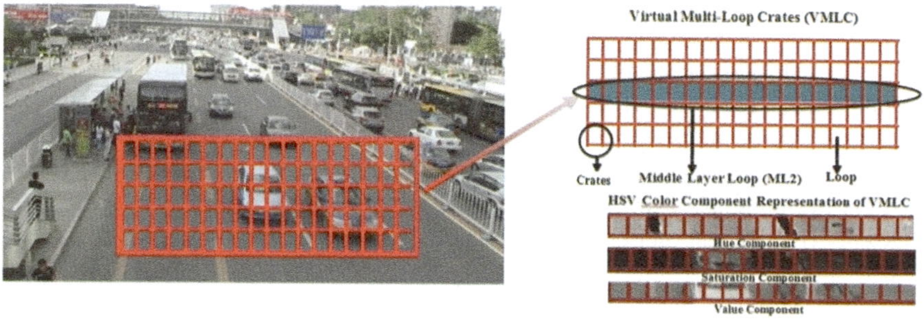

Fig. 2 Virtual multi-loop crates (VMLC)

primarily operate on specific segments or directions of movement rather than the entire content of the traffic video.

Significance of VMLC The beneficial features of the proposed VMLC detection framework are (1) reduction of computational space: Image processing techniques are generally preformed at pixel level. When ML2 is considered instead of processing the entire image, it reduces the number of pixels for processing. (2) A complicated video content may become simple content: VMLC removes the surrounding disturbances and movements, which reduces the complex detection into a simple or static background detection. (3) Reduced false positives: Using VMLC ensures to eliminate unwanted and complicating disturbances that subsequently elevate the accuracy of vehicle detection with reduced false positives. (4) Ease of use: VMLC framework can be easily integrated with any image processing technique using video-based vehicle detection application with simple recurrent detection process for all the crates. (5) Traffic information extraction: Apart from vehicle detection, the proposed framework also aids in classifying vehicles and detecting speed of the moving vehicles.

3.2 *Concurrent Spatial Color Information Processing System (CSCIPS)*

The real world traffic scene is represented virtually in the form of image that needs to be preserved. Almost all the existing vehicle detection process diminishes this information by losing spatial color information by suppressing three levels of spatial domain knowledge into single level that has substantial features and varying illumination information. Thus, the proposed framework further improves the detection accuracy by preserving this spatial information by using Hue, Saturation, and Value (HSV color information) elements of the image. The processing of each spatial domain is done concurrently making it similar to a gray-level image processing as represented in Fig. 2. It has been identified that images in HSV color model show significant difference in lighting changes required to identify the moving object in the outdoor image sequence [13, 14]. The enhancements using concurrent processing spatial color information for classic motion-based Background Subtraction (BS) methods including running average and Mixture of Gaussian (MoG) are presented in this section. In any BS, the foremost challenge is to define the background image for comparison. The Updated Background Image (UBI) of running average for C color components where $C \in H, S, V$ of the Pixel Intensity (PI) at location (a, b) for a crate of size M×N in [4, 5, 13, 14] is shown in Eqs. 1 and 2.

$$AI_C = 1/B \sum_{j=1}^{B} (a, b)_j \tag{1}$$

$$(UBI_C)_i = \alpha(UBF_C)_{i-1} + (1 - \alpha)AI_C \tag{2}$$

where AI is the Average of B images of ith iteration for Cth color component, UBI_C is the updated background image for Cth color component, and α is the weightage parameter. For MOG, the background image modeling using Gaussian distribution as in [17] is given in Eqs. 3 and 4.

$$F(PI_{T,C}) = \sum_{j=1}^{K}(\omega_j, C, T)\ \delta C((PI_{T,C}), (\mu_j, T, C), (\epsilon_j, T, C)) \tag{3}$$

$$\delta C((PI_{T,C}), (\mu_j, T, C), (\epsilon_j, T, C)) = \frac{exp(-1/2(PI_{T,C} - (\mu_j, T, C))^T \epsilon_j, T, C(PI_{T,C} - (\mu_j, T, C))}{(2\Pi^{D/2}\epsilon_j, T, C)^{1/2}} \tag{4}$$

where $F(PI_{T,C})$ represents the K Gaussian distribution for pixel intensity at time T for Cth color component, δC, μ, and ϵ represent the probability of observing the current pixel, mean, and variance, respectively, and (ω_j, C, T) is the jth mixture of Gaussian model weightage. The Gaussian models are updated using a Mahalanobis or log-likelihood distance using Eq. 5. The Cth component pixel intensity corresponding to K mixture models represents the background pixel $BP_{j,C}$ using threshold TH = 2.5–3. The rest of the pixels correspond to the moving object and given value 0. The weights ω are updated using learning rate η in Eqs. 6 and 7.

$$S((PI_{T,C}), (\mu_j, T, C), (\epsilon_j, T, C)) = \begin{cases} BP_{j,C}\ if((PIT, C) - (\mu_j, T, C) < TH(\epsilon_j, T, C)) \\ 0 \qquad Otherwise \end{cases} \tag{5}$$

$$\omega_j, C, T = \eta\ \theta((\omega_j, C, T-1), (PI_{T-1,C})) + (1-\eta)\omega_j, C, T \tag{6}$$

$$\theta((\omega_j, C, T-1), (PI_{T-1,C}) = \begin{cases} 1\ if(S((PI_{T,C}), (\mu_j, T, C), (\epsilon_j, T, C)) = BP_{j,C}) \\ 0 \qquad Otherwise \end{cases} \tag{7}$$

Using the above set of equations, the BS method is carried out for each crate of VMLC. Also, the above set of equations shows that the proposed framework with spatial information can be integrated and implemented concurrently in any vehicle detection algorithm.

4 Performance Evaluation

During different environmental conditions and traffic, videos taken at various camera angles were considered for vehicle detection using CSCIPS with VMLC. The benchmark datasets such as DETRAC and self-recorded videos similar to the screenshot of video in Fig. 1 were used for detection [4]. The videos were of different resolutions such as 800 × 600, 1024 × 576, 1280 × 720. The implementations are done using Visual Studio C++ programming language with OpenCV and OpenMP library files [18]. The classic Background Subtraction (BS) algorithms of K-Nearest Neighbor (KNN), Mixture of Gaussian (MOG), and running average are used for analysis [5, 6, 15, 16]. The screenshot of entire image vehicle detection using CSCIPS for KNN, MOG, and running average is shown in Fig. 3 for visual interpretation. Other algorithms and state-of-the-art detection techniques are difficult to model using their respective main function and operating parameters; hence, easily available methods were chosen [18]. It should be noted that the performance of detection using VMLC with CSCIPS is analyzed and not the detection algorithms. The detection of vehicles using VMLC is analyzed using detection rate. The rate of detection determines the number of objects identified by the algorithm but not specifically the vehicles. The detection rate for the proposed VMLC and other existing detection frameworks such as selected region, entire image as grids, and entire image is analyzed. For easy analysis of performance, algorithms are tested with three types of traffic composition: low, medium, and high

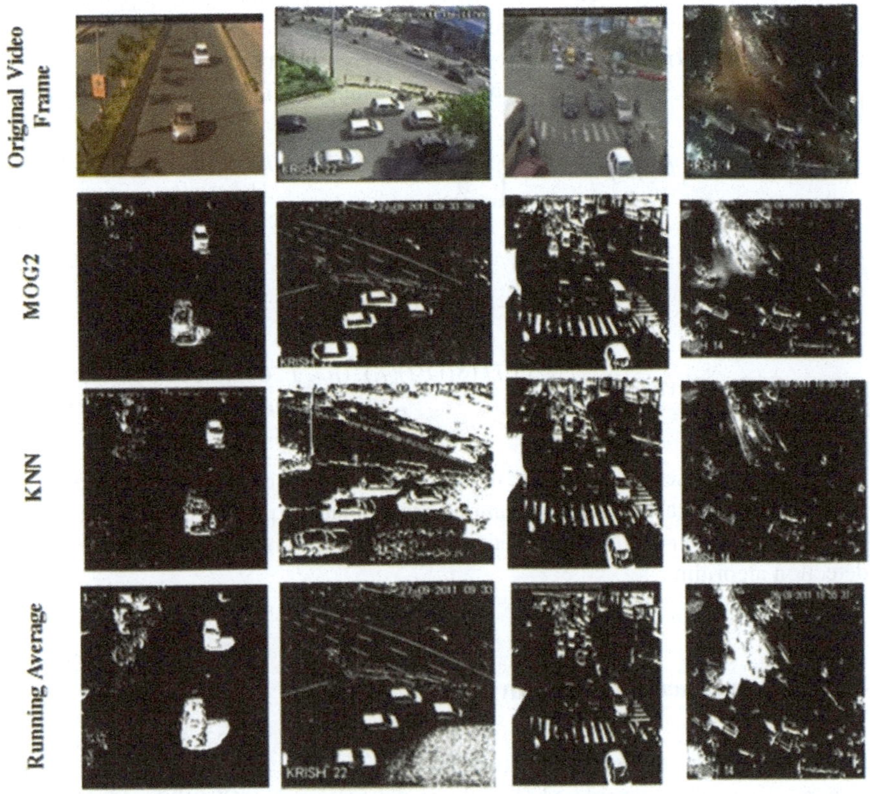

Fig. 3 Results for detection algorithms using CSCIPS

(vehicle density <40, 40–120, and >120, respectively). Figure 4a–c shows detection ratio for few observations in traffic videos with composite traffic graded as low, medium, and high traffic, respectively. It is seen from the results of detection rate that the detection of moving object from the scene is very high for entire image as the number of pixels is high and the region of coverage is also found to be high

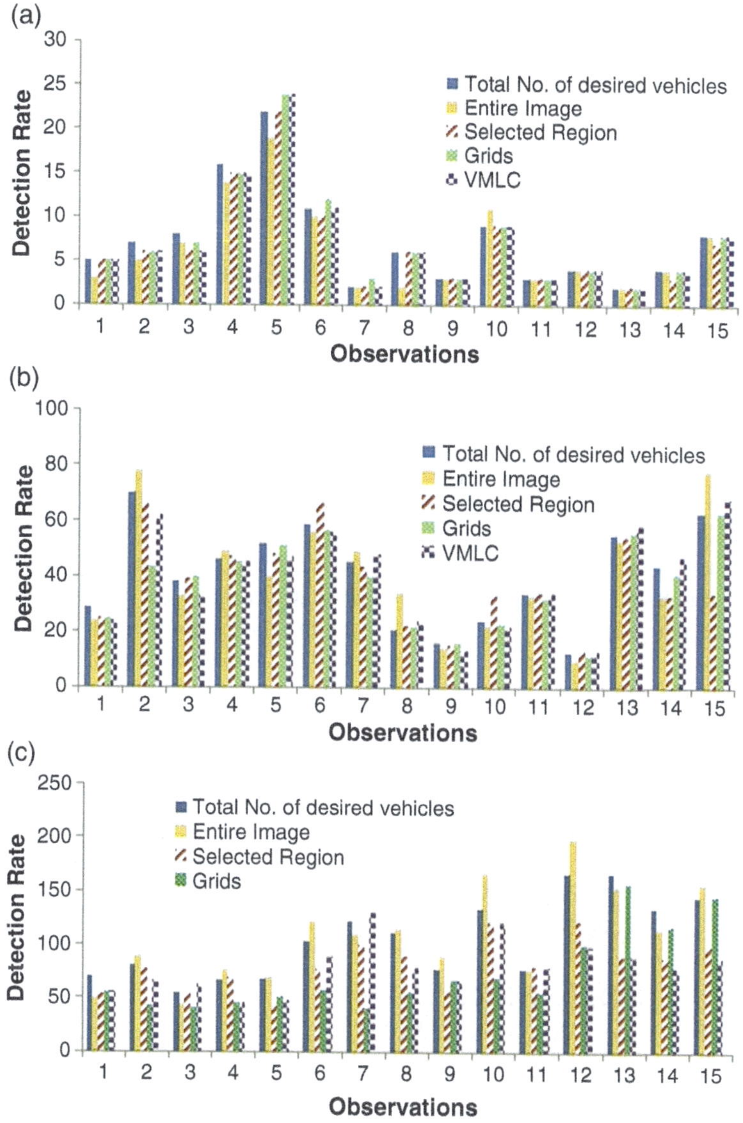

Fig. 4 Detection rates for VMLC. (**a**) Detection rates for framework in low traffic. (**b**) Detection rates for framework in medium traffic. (**c**) Detection rates for framework in high traffic

compared to other frameworks. The detection of unwanted objects in the scene makes the detection ratio higher than the desired number of vehicles, particularly in high traffic conditions as in Fig. 4c.

The detection of moving objects is therefore more or less similar for selected region, grids, and VMLC and less relative to the entire image detection process. The overall detection rates from different benchmark traffic videos using VMLC with CSCIPS are shown in Fig. 5. It is also seen here that the average moving object detection for an entire image is very high, which does not mean that it has accurately detected vehicles. The overall false positive rates for the frameworks in static and dynamic background are shown in Fig. 6a, b, respectively. It is apparent that the entire image-based processing is particularly vulnerable to error detection and high false positives. The false positives in selected area, grids, and VMLC are significantly reduced. The proposed framework for the detection thus indicates the lowest false positive among all leading to improved and efficient accuracy in vehicle detection.

The precision–recall curves for different frameworks show the accuracy of detecting the desired vehicle from the candidate detection process in Fig. 7. Due to the increased false positive rates, the entire image-based processing shows the least accuracy. The framework of VMLC reduces the unwanted detection of object and also reduces the false positives drastically especially in a highly dynamic background thereby increasing detection accuracy. As proposed in Sect. 3.2 VMLC uses the color info in HSV, which increases the detection accuracy further. The detection process in VMLC with grayscale image and HSV color model is also presented in Fig. 7. It is seen that the VMLC using HSV color model helps in increasing the accuracy of detection when compared with the grayscale image. The performance of VMLC is further analyzed based on the number of Frames Processed per Second (FPS) and the memory or storage space required to process the videos for vehicle detection. It is seen from Fig. 8 that the time taken to process

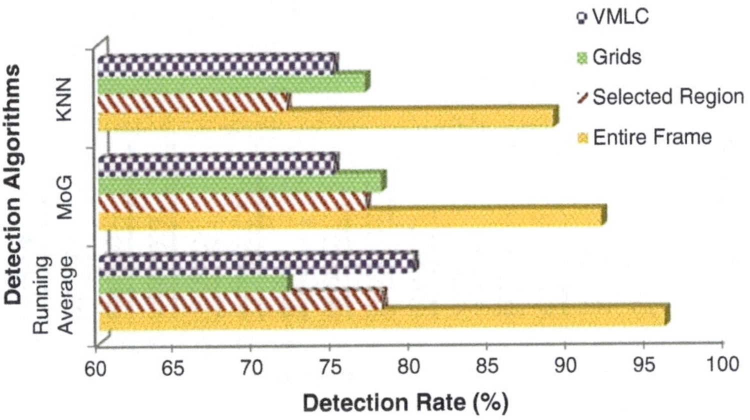

Fig. 5 Average detection rates for framework

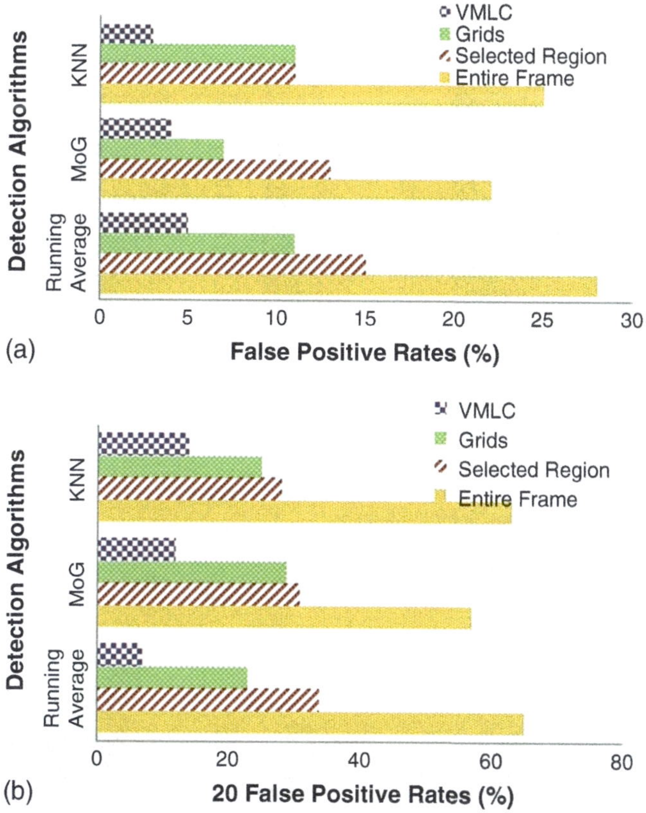

Fig. 6 False positive rates for VMLC. (**a**) False positive rates for static background video. (**b**) False positive rates for dynamic background video

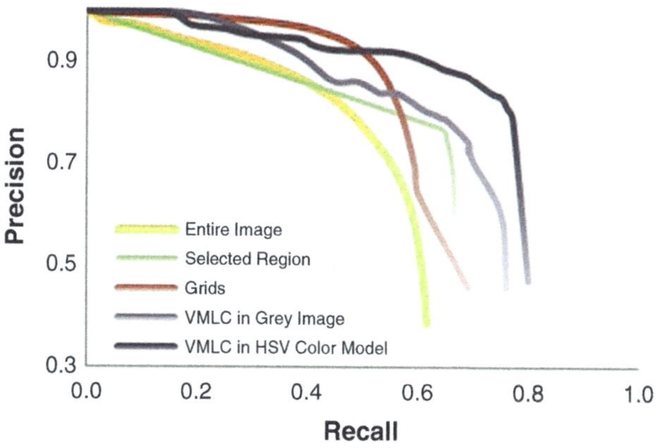

Fig. 7 Precision and recall curves for framework

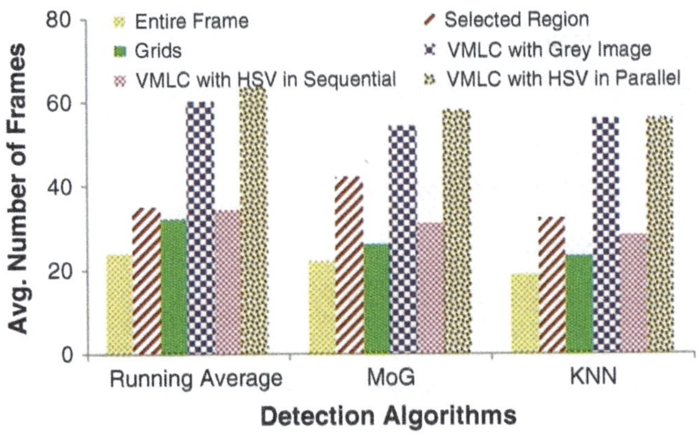

Fig. 8 Average FPS for framework

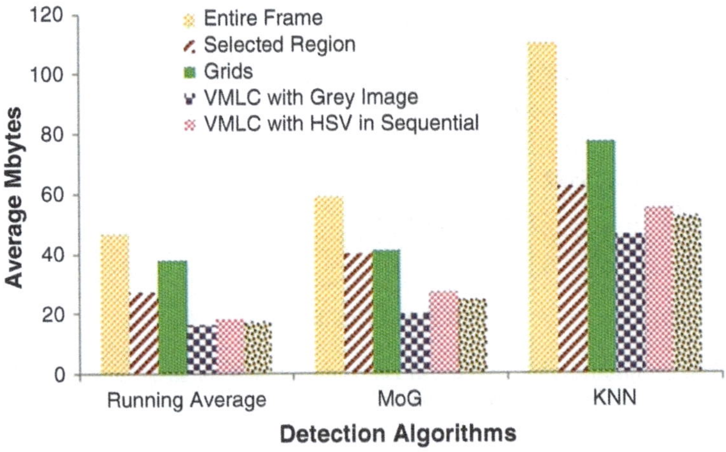

Fig. 9 Storage space for framework

the frame using VMLC is very low due to the drastic reduction in the total number of pixel for processing,which improves the number of FPS. It is also seen that the processing time using a grayscale image and a parallel HSV color model is similar. The average memory required for storage during vehicle detection using VMLC is the least compared to other detection frameworks as shown in Fig. 9. Also the space required for processing HSV color model in sequential and parallel is almost similar. To sum up VMLC performance, the results show that any video-based application for vehicle detection can incorporate it into its module. It also reveals that multi-loop structure is novel and can be used in any direction of motion and camera angle of traffic video. Due to the drastic reduction of the number of pixels in the process, VMLC shows reduced processing time per frame and storage space. It is also verified that improved accuracy with minimal false positives of vehicle detection is achieved using VMLC frameworks with spatial HSV color information processed concurrently.

5 Conclusion

The robustness of any vehicle detection process of Intelligent Transportation System (ITS) application based on computer vision depends on the image processing techniques employed. This paper proposes a novel vehicle detection structure known as Virtual Multi-Loop Crates (VMLC) to easily integrate into any video-based vehicle detection application, improve detection accuracy with enhanced processing time, and reduce storage space. The accuracy of detection is further improved by using Concurrent Spatial Color Information Processing System (CSCIPS) that ensures to process all the color information without suppressing it. The performance of the proposed VMLC with CSCIPS is analyzed by the detection rates of Background Subtraction (BS) algorithms in diverse benchmark traffic videos. It is seen from the results that VMLC with concurrent processing of HSV color model reduces the false positives, processing time, and storage space compared to other frameworks, resulting in improved vehicle detection process accuracy.

Acknowledgments All the research work related to this publication is supported by Centre of Excellence in Transportation Engineering (CETransE), National Institute of Technology, Tiruchirappalli, Tamil Nadu, India, which is approved and funded by Ministry of Human Resource Development (MHRD), Government of India.

References

1. Yang, Z., & Pun-Cheng, L. S. C. (2018). Vehicle detection in intelligent transportation systems and its applications under varying environments: A review. *Journal on Image and Vision Computing, 69*, 143–154.
2. Boukerche, A., Siddiqui, A. J., & Mammeri, A. (2017). Automated vehicle detection and classification: Model, methods and techniques. *ACM Journal on Computing Surveys, 50*(5), 1–39.
3. Manipriya, S., Mala, C., & Mathew, S. (2020). Significance of real time systems in intelligent transportation systems. In *Chapter 5 in handling priority inversion in time-constrained distributed databases* (pp. 61–85). IGI Global Publications.
4. Wen, L., Du, D., Cai, Z., Lei, Z., Chang, M.-C., Qi, H., Lim, J., Yang, M.-H., & Lyu, S. (2020). UA-DETRAC: A new benchmark and protocol for multi-object detection and tracking. *Journal on Computer Vision and Image Understanding, 193*, 1–27.
5. Manipriya, S., Mala, C., & Mathew, S. (2020). Virtual mono-layered continuous containers for vehicle detection applications in intelligent transportation systems. *Journal of Discrete Mathematical Sciences and Cryptography, 23*(1), 321–328.
6. Bouwmans, T., & Garcia-Garcia, B. (2019). Background subtraction in real applications: Challenges, current models and future directions. Eprint 1901.03577.

7. Seenouvong, N., Watchareeruetai, U., Nuthong, C., Khongsomboon, K., & Ohnishi, N. (2016). Vehicle detection and classification system based on virtual detection zone. In *International Joint Conference on Computer Science and Software Engineering* (pp. 1–5).
8. Induja, C. (2018). A survey on vehicle counting—double virtual line. *Journal for Research in Applied Science and Engineering Technology, 6*, 4471–4473.
9. Chen, L., Ye, F., Ruan, Y., Fan, H., & Chen, Q. (2018). An algorithm for highway vehicle detection based on convolutional neural network. *EURASIP Journal on Image and Video Processing, 109*, 1–7.
10. Sun, W., Zhang, X., Shi, S., He, J., & Jin, Y. (2017). Vehicle type recognition combining global and local features via two-stage classification. In *Mathematical Problems in Engineering*.
11. Pae, D. S., Choi, I. H., Kang, T. K., & Lim, M. T. (2018). Vehicle detection framework for challenging lighting driving environment based on feature fusion method using adaptive neuro-fuzzy inference system. *International Journal of Advanced Robotic Systems* 1–15.
12. Mohamed, A., Issam, A., Mohamed, B., & Abdellatif, B. (2015). Real-time detection of vehicles using the haar-like features and artificial neuron networks. *Advanced Wireless, Information, and Communication Technologies, 73*, 24–31.
13. Manipriya, S., Mala, C., & Mathew, S. (2014). Performance analysis of spatial color information for object detection using background subtraction. *IERI Procedia, 10*, 63–69.
14. Manipriya, S., Ramadurai, G., & Reddy, V. B. (2015). Grid-based real time image processing (GRIP) algorithm for heterogeneous traffic. In *International Conference on Communication Systems and Networks* (pp. 1–6).
15. Piccardi, M. (2004). Background subtraction techniques: A review. In *IEEE International Conference on Systems, Man and Cybernetics* (Vol. 4, pp. 3099–3104).
16. Shah, M., Deng, J. D., & Woodford, B. J. (2014). Video background modelling: Recent approaches, issues and our proposed techniques. *Springer Journal on Machine Vision and Applications, 25*, 1105–119.
17. Stauffer, C., & Grimson, W.E.L. (1999). Adaptive background mixture models for real-time tracking. In *Proceedings of CVPR* (pp. 246–252).
18. Background Subtraction using OpenCV (2019). https://docs.opencv.org/3.4/d1/dc5/tutorial_background_subtraction.html

Efficient Routing Strategies for Energy Management in Wireless Sensor Network

Raviteja Kocherla, M. Chandra Sekhar, and Ramesh Vatambeti

Abstract Wireless Sensor Network (WSN) refers to a group of distributed sensors that are used to examine and record the physical circumstances of the environment and coordinate the collected data at the centre of the location. This WSN plays a significant role in providing the needs of routing protocols. One of the important aspects of routing protocol in accordance with Wireless Sensor Network is that they should be efficient in the consumption of energy and have a prolonged life for the network. In modern times, routing protocol, which is efficient in energy consumption, is used for Wireless Sensor Network. The routing protocol that is efficient in energy consumption is categorized into four main steps: CM – Communication Model, Reliable Routing, Topology-Based Routing and NS – Network Structure. The network structure can be further classified as flat/hierarchical. The communication model can be further classified as query, coherent/non-coherent, negotiation-based routing protocol system. The topology-based protocol can be further classified as mobile or location-based. Reliable routing can be further classified as QoS (Quality of service) or multiple-path based. A survey on routing protocol that is energy-efficient on Wireless Sensor Network has also been provided in this research.

Keywords Routing protocols · Energy efficiency · Wireless sensor networks

---

R. Kocherla (✉)
Research scholar, Department of CSE, Presidency University, Bangalore, India

M. Chandra Sekhar
Department of CSE, Presidency University, Bangalore, India
e-mail: chandrasekhar@presidencyuniversity.in

R. Vatambeti
Department of CSE, CHRIST (Deemed to be University), Bangalore, India

A. Haldorai et al. (eds.), *3rd EAI International Conference on Big Data Innovation for Sustainable Cognitive Computing*, EAI/Springer Innovations in Communication and Computing, https://doi.org/10.1007/978-3-030-78750-9_6

1 Introduction

Internet of Things (IoT) is the ecology of interconnected objects and devices that transmit and receive data via Internet. It is an imperceptible and smart network that communicates with each other by employing embedded technology that enables them to sense and control under diverse conditions. The IoT grants immediate access with high efficiency and productivity to the information related to any object or device. Wireless sensor networks (WSNs) operate as an intermediate that connects the real world and virtual digital world. These are highly scattered networks of lighweight, small, battery-embedded sensor nodes connected to each other and are accountable for sensing and transmitting information to the Internet. WSN is a very attractive area for numerous applications, which provides the most challenging solutions, such as battlefield monitoring and military reconnaissance, tsunami detection, patient health monitoring, disaster surveillance and crises management, habitat monitoring and industrial automation. Sensor nodes (SNs) are comprised of memory, processors, sensing elements, batteries and transceiver. These sensor nodes can be disposed in large amount in sensing field that generates a massive amount of data and sends it to the base station for further processing. Despite this, due to the tiny nature of sensor nodes, it has a few limitations such as memory, bandwidth, processing capability and battery power. It could not be argued that energy management is the first concern when developing a reliable and efficient WSN.

Usually, when WSN is used for remote monitoring of the environment, a long existence of the sensor network guarantees the overall efficiency and reliability of the monitoring system, and it also maintains a great deal of human effort to preserve the sensor network functionality. By recognizing all the factors contributing to the consumption of energy to sustain all operations within the WSN, usually a huge proportion of energy is used during communication. By minimizing the transmitting hop count, their average length could minimize the consumption of energy. Ultimately, the WSN's lifespan will be effectively extended as anticipated. In WSN, the collection of data from the sensor networks can be event-based, periodic or query-based. In periodic data gathering, at regular intervals, the sensor nodes gather data by sensing the environment and transmit the sensed data to the base station. The routing approach acts a vital role in enhancing the lifespan of the network. Most of the routing protocols in WSN are divided into different classes, such as flat, location-based and hierarchical-based techniques. Many energy-efficient routing protocols have been suggested to solve the early energy dissipation problem and likely to succeed.

Hierarchical routing in sensors networks have three different kinds of routing based on their communication with the base station (BS) or sink, that is, tree-based approach, cluster-based approach and chain-based approach. The cluster-based approach is a key contestant for improving the network lifespan. Cluster-based routing has attained this with lowest overhead; the cluster formation is effective approach for data collection, where the sensor nodes are grouped into separated regions. Every group has a leader node in cluster, known as Cluster Head (CH), which has the responsibility to communicate with cluster members (CMs) and collect data from them. One vital role of the CH is data aggregation, where it eradicates the unnecessary and redundant data packets, thus minimizing the delay and overhead. The clustering lessens the routing complexity by minimizing the number of nodes and size in the routing pathway, which strengthen the efficient utilization of memory and bandwidth. In addition, it accomplishes the collision-free access channel among the CM and CH by utilizing the Time Division Multiple Access (TDMA) schedule effectively. Thereby, it enhances the sensor network's scalability and stability. The Received Signal Strength (RSS) plays a very important role to accomplish self-configuration characteristics in WSNs, where every node itself finds the neighbour nodes. Furthermore, it determines the distance to the neighbours and link quality with the intentions that it effectively constructs the reliable network with strong link quality by avoiding collision, interference, packet retransmission and overhearing. Therefore, it remarkably saves battery power and enhances the network lifetime.

In addition, it greatly minimizes the involved number of men and hardware complexity in network setup. The unequal clustering techniques and its benefits were discussed in, where the cluster size and CH selection was investigated on the basis from distance to the BS. Furthermore, it tackles the early depletion of battery and bottleneck problem nearer to the base station. The CH selection using distributed approaches has been proposed, which enhances the network lifetime. LEACH algorithm enhances the network lifespan, the cluster head is selected on random basis, which does not guarantee appropriate distribution of nodes and optimal results. The sensor nodes with low and higher residual energy have equal opportunities to become the CH of the group. Therefore, the selection of low residual energy node as CH resulting to die quickly and hence shortens the lifespan of the network. Load Balancing Protocol (EESAA) considers the parameters such as network lifetime, stability period and throughput for WSNs by examining and enhancing the performance of clustering algorithm. Cluster heads receive data from the sensor nodes in the same cluster, fuse it and send it to the base station. In the entire network, EESAA does not guarantee any uniform distribution of the CHs. A Cluster-Chain Mobile Agent Routing (CCMAR) Algorithm is proposed by [1], where the CH selection value of each node is evaluated on the basis of the node's residual energy, the number of bits transmitted, distance from the neighbour node and amount of energy required for the transmission of one bit. Despite this, the CCMAR algorithm leads to the isolated node problem, owing to the fact that it is unaware of the selection of CH value of the neighbours' neighbour because the nodes construct a chain and send data to its nearest neighbour and by this way reach CH. Furthermore, there should be high delay when each node sends and receives data from its neighbour (Figs. 1 and 2).

An advanced algorithm is presented known as R-LEACH; this algorithm seeks to elect a CH by taking into account the most important parameters such as initial energy, the individual sensor node's remaining energy, and the optimal number of CHs in the network. The amendment is carried out in classical LEACH (Low Energy

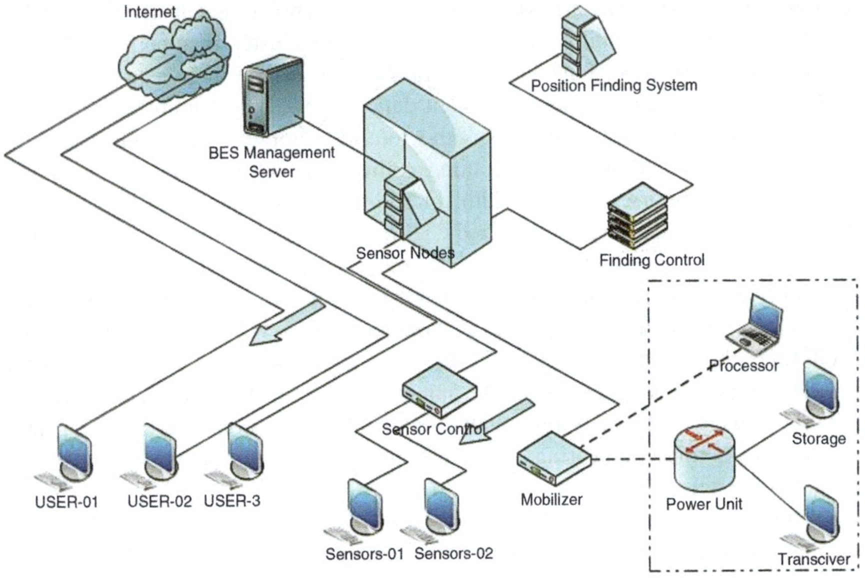

Fig. 1 Outline of wireless sensor network

Adaptive Clustering Hierarchy) algorithm. Following the completion of each round, the non-CH nodes' residual energies are examined; for the current around the node with better residual energy has a higher chance for CH selection. This strategy would thwart the early depletion of energy nodes and thus prolong the network life span. The optimal cluster formation and communication with BS using shortest path (i.e. minimum hop-count) are very important parameters for enhancing the network lifetime, which are not considered by R-LEACH. Despite this, R-LEACH only considers the CH selection based on its residual energy but do not take into account the shortest distance from BS, distance of neighbour nodes and number of neighbour nodes as input parameters (Fig. 3).

2 Routing Protocols System in Wireless Sensor Network

In this segment, an analysis was conducted based on routing protocol with state of the art in Wireless Sensor Network. The routing protocol in Wireless Sensor Network can be classified as a flat, hierarchical, location-based model, which is based on their network structuring. The nodes of the flat-based routing protocol are assigned equally by means of their function as well as roles. The nodes of the hierarchical-based routing protocol will have various roles to perform in the networking system. The nodes of the location-based routing protocol are exploited by the routing data in the system. The routing protocol is very adaptive in nature;

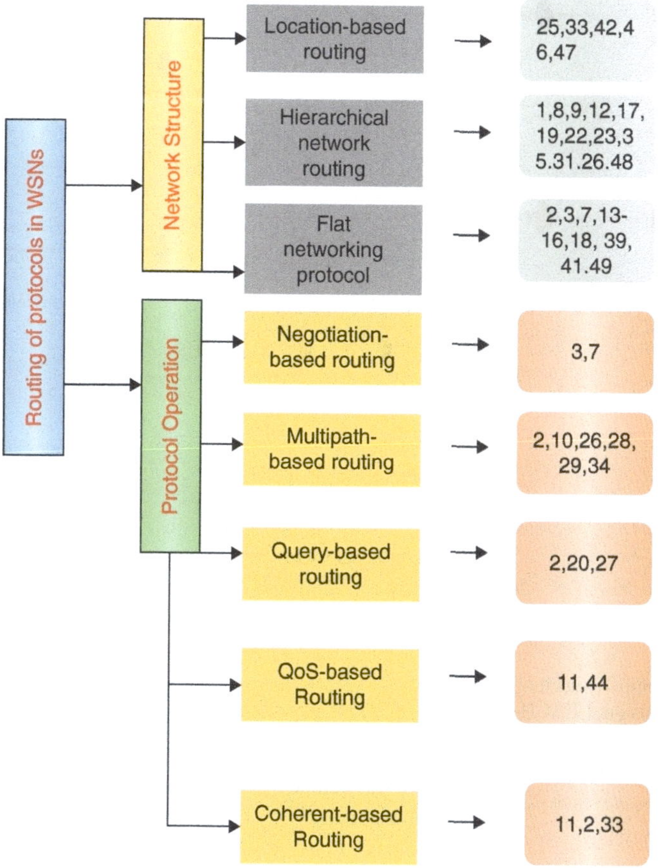

Fig. 2 Routing protocols in WSNs: a taxonomy

this may happen if the parameters control themselves in order to work accordingly for the conditions of the present network along with sufficient energy. The routing protocol is further divided into three groups as reactive routing, proactive routing, and hybrid routing based on the destination identified by the routing system. The routes of the proactive system are computed before the requirement, but in the case of a reactive, the route is computed only if it is needed; in a hybrid system route is computed either before or after demand.

If the nodes are static table-driven protocol, they are preferred than the reactive routing protocol. A sufficient level of energy was consumed for locating the route of the nodes and setting up of reactive routing protocol. Cooperative routing is another type of routing protocol. In this protocol, the data are sent to the central node where the data can be combined to use it for further process, by which routing cost can be reduced in the concept of usage of energy. Various types of routing protocols are also addressed in this research chapter. These protocols are classified based on

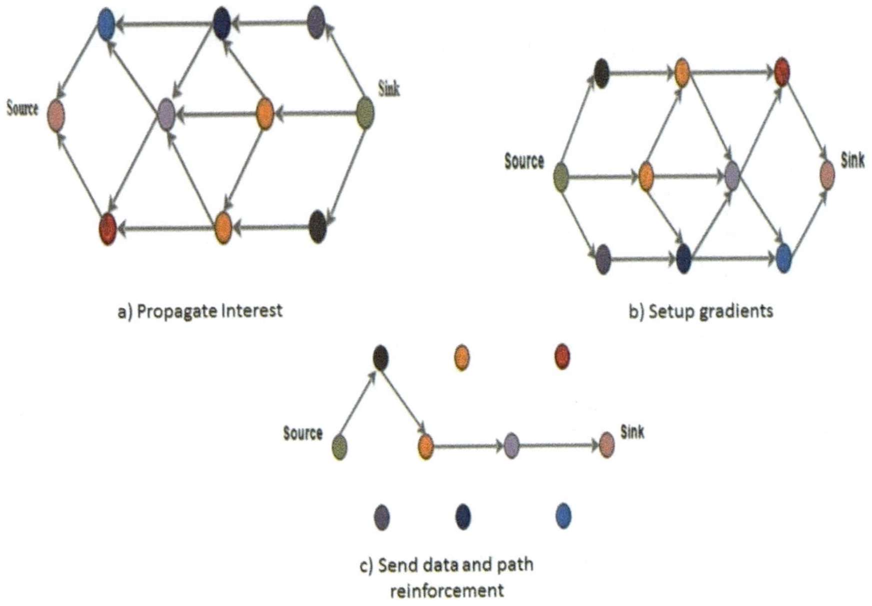

Fig. 3 Sensor network based on interest diffusion

Table 1 Comparison among directed diffusion, LEACH and SPIN

	SPIN	LEACH	Directed diffusion
Usage of metadata	1	0	0
Resource awareness	1	1	1
Network lifetime	4	5	4
Optimal route	0	0	1

the structure of network and operating system of protocols. The classification of protocol is illustrated in Fig. 2.

Various studies have examined this subject involving the development and implementation of different energy-efficient routing protocols [2]. Specifically, the chapter seeks to unearth issues such as the methodologies adopted, the proposed energy-efficient routing models in different scenarios and applications, and the scholars' results regarding the impact of the proposed algorithms on the network lifetime of WSNs. From the results, a common trend and emerging theme that have been established demonstrate that when energy-efficient routing systems are implemented, they are likely to reduce the energy consumption of WSN networks. With energy consumption minimized, the resultant efficiency with which the energy-efficient routing protocols are associated contributes to an increase in the network lifetime (Table 1 and Fig. 4).

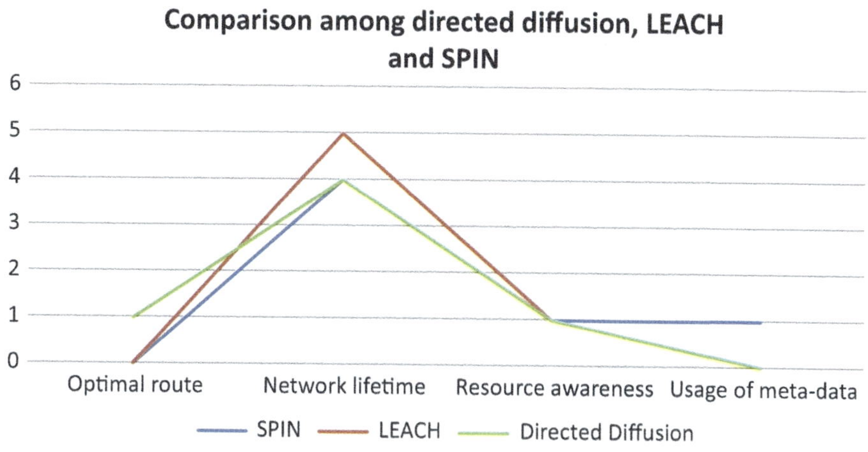

Fig. 4 Comparison among directed diffusion, LEACH and SPIN *Note 0-No, 1-Yes, 5-Very good, 4-Good

3 Network-Structure-Based Protocols

The protocols that come under Network-based routing protocol will be elaborated in this section. This protocol works significantly in the functioning process of routing protocol in Wireless Sensor Network.

3.1 Flat Routing

The first classification of flat routing protocol is known as multi-hop. All the nodes of this flat protocol will do similar job, and sensor nodes are used for performing the sensing task. In this system, more nodes are involved in the activity, so it may be difficult to assign global identifying nodes for each node. The TEEN and APTEEN protocols are motivated to develop same kinds of other protocol. The advantages of these two protocols have been elaborated and discussed in the upcoming section as follows (Fig. 5).

3.2 Sensor Protocols for Information via Negotiation

[3] proposed a framework called the SPIN (Sensor Protocols for Information via Negotiation). This protocol comes in different varieties and is designed for addressing the deficiency of flooding and gossiping. It is highly different when compared to the conventional data-centric protocol. In this protocol, the user can

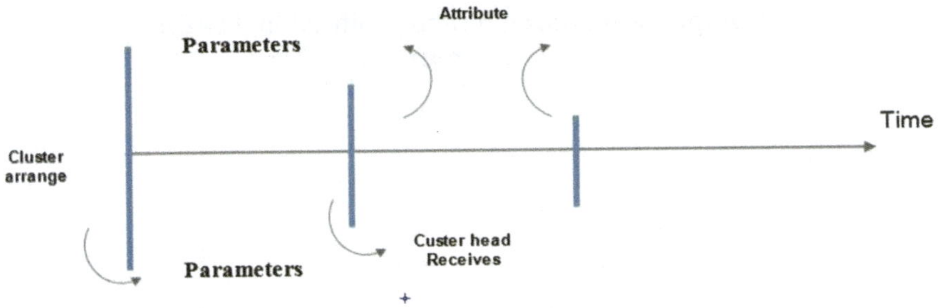

a) TEEN

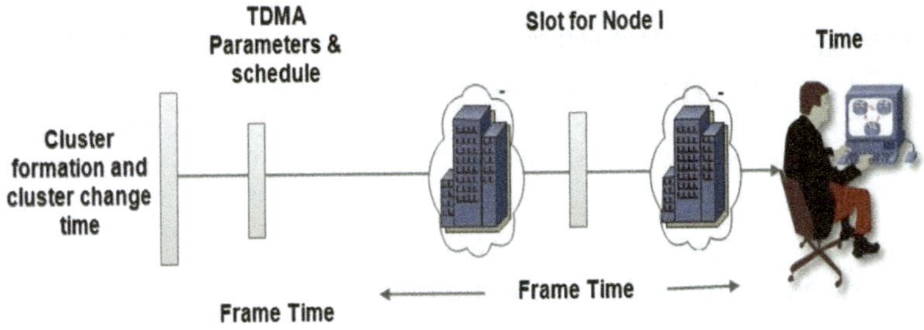

b) APTEEN.

Fig. 5 Timeline for the operation

query any nodes in the system and get information from them. This protocol uses the data from the nodes those are nearer to it and those with similar data so only data distribution will be the further process in it. The Sensor Protocols for Information via Negotiation uses the current energy and runs the network, and the working processes are done with energy remaining in the system. The working structure of this protocol is based on time-driven pattern through which the information is distributed to the entire network, even if the information is not needed for the users. The main aim of SPIN protocol is to address the deficiency in classic flooding through the process of negotiation and adaption of resource.

The SPIN protocol is designed with the idea of two basic features:

1. Sensor nodes work efficiently and help in conserving more energy by sending data only required by the sensor instead of sending all data.
2. There is some traditional protocol called gossiping or flooding based on the usage of energy, and more bandwidth is needed while sending data or extra data for the network system.

The important criticism of flooding-based routing protocol is the occurrence of sudden collapse in the networking system, which is due to identical message sent to the same nodes twice or more number of times. Overlapping process in the system occurs when two nodes of similar region send same packets to the similar neighbourhood node [4].

In gossiping-based routing protocol, the problem of sudden collapse in the networking system does not occur blindly; it can be avoided by choosing random nodes and sending packets to different neighbourhood nodes. The propagation of data in this protocol takes more time.

The other important protocols of SPIN family includes

- SPIN-BC protocol: It is utilized for broadcasting the channels in the network
- SPIN-PP protocol: It is utilized for having point-to-point (direct) communication
- SPIN-EC protocol: The working of this protocol is similar to that of the above-mentioned protocol but with question of energy involved in it.
- SPIN-RL protocol: If a channel is lossy, this protocol helps in adding to fill the space of lossy channel.

3.3 Directed Diffusion (DD)

[5] proposed a framework utilizing DD (directed diffusion) in wireless sensor network. The DD is the DC (data-centric) paradigm that all the nodes are named under the attribute value. The DC is used for combining different nodes together, which are obtained through different sources by avoiding redundancy, reducing the transmissions, through which the energy will be saved and will be prolonged for long time for running the network system. The present DC allows data to enter from various sources; it is comparatively different to that of the conventional one. In the DD system, the sensor is capable of measuring events through which information is created with the help of the neighbourhood node. BS data are processed in the interest of the broadcasting channel. Interest here is referred to the task needed to be performed by the network. An interest spreads over a wider area in the network by hop-by-hop and later broadcasted to all nodes in the neighbourhood. The gradients are responsible for deciding the attribute value and direction. The gradients' strength varies from neighbour to neighbour through which the flow of information also varies gradually. The loops will not be checked at this time but will be considered at a later period of time. Fig. 3 illustrates the working principle of DD.

DD in SPIN varies in two aspects:

1. In DD the queries in the data are revealed with the request made by BS, where the BS sends all these queries to the nodes by flooding the task. In SPIN only nodes, those who are interested can query the data.
2. The communication pattern of DD is within neighbourhood nodes which are capable of performing data aggregation and caching.

With the help of the SPIN, the need for maintaining global network topology is not required in DD system. The DD will not work efficiently in the network, which asks for continuous delivery of data to the BS. This is due to the reason that queries are functioned only if there is demand for it. To solve queries, the data must be matched in order to execute it.

3.4 Rumour Routing

Rumour routing is a WSN routing-based protocol, which aims at lower energy consumption than other protocols that flood the whole network with varies types of queries or event messages. But in some cases, the nodes only need small amount of data; so the idea of flooding is not necessary in this situation[6]. In some situations, the total amount of event may be smaller and the amount of queries will be larger. The main aim of this rumour routing is to match the queries with the correct nodes by which unnecessary dumping of data can be avoided. The rumour routing has employed Agent to deal with flooding event in the networking system. If event is detected by the nodes, then it will be added in the table with the category called the event table and there the agents are generated. The agent in the network travel to pass information about events to the nodes that are located in a far distant location. When event generates a query, the nodes which knows to route to reach the query will reach as soon as possible by knowing the information from the event table. The flooding process will not occur in this rumour routing as well the cost for communication will also be manageable. In this process, there is only one route between the sources to reach the destination, but in the case of DD it is provided with multiple route at low cost. The rumour routing is capable of working with significant energy consumption when compared with event flooding process and is also capable of handling failure in the nodes. The rumour routing works with lesser amount of events. With a greater number of events, the working speed and maintenance of nodes in the system as well as table becomes less efficient (Table 2).

3.5 Minimum Cost Forwarding Algorithm (MCFA)

The MCFA computes the least cost from each node to the base station (BS). If the node is in shortest route, the data will be retransmitted, the similar procedure continues until the packet reaches the BS. The message forwarded to the nodes will be broadcasted for the other nodes in the neighbourhood. The message received by the node will first check the cost to reach the other nodes and the base station.

In MCFA, the nodes must know the shortest path to reach the other nodes and the BS. This is achieved with the following process:

- In base station, the messages will be broadcasted with 0 cost, while the node sets it cost to be ∞ in the base station.
- Each node after receiving the message from base station will check the estimation cost whether it is lower to higher than the estimation.

Table 2 Classification and comparison of routing protocols in wireless sensor networks

	Classification	Query-based	Multipath	Scalability	State complexity	QoS	Localization	Data aggregation	Negotiation-based	Power usage	Position awareness	Mobility
SPIN	Flat	1	1	2	1.5	No	No	1	1	2	0	Possible
Directed Diffusion		1	1	2	1.5	0	1	1	1	2	0	2
Rumour Routing		1	0	4	1.5	0	0	1	0	N/A	0	5
GBR		1	0	2	1.5	0	0	1	0	N/A	0	2
MCFA		0	0	4	1.5	0	0	0	0	N/A	0	0
CADR		0	0	2	1.5	0	0	1	0	2	0	0
COUGAR		1	0	2	1.5	0	0	1	0	2	0	0
ACQUIRE		1	0	2	1.5	0	0	1	0	N/A	0	2
EAR		1	0	2	1.5	0		0	0	N/A	0	2
LEACH	Hierarchical	0	0	4	CHs	0	1	1	0	Maximum	0	Fixed BS
TEEN & APTEEN		0	0	4	CHs	0	1	1	0		0	
PEGASIS		0	0	4	1.5	0	1	0	0		0	
MECN & SMECN		0	0	1.5	1.5	0	0	0	0		0	0
SOP		0	0	1.5	1.5	0	0	0	0	N/A	0	0
HPAR		0	0	4	1.5	0	0	0	0		0	0
VGA		0	1	4	CHs	0	1	1	1		0	0
Sensor aggregate		Possible	0	4	1.5	0	0	1	0		0	2
TTDD		Possible	Possible	1.5	3	0	0	0	0	2	1	1
GAF	Location	0	0	4	1.5	0	0	0	0	2	0	2
GEAR		0	0	2	1.5	0	0	0	0	2	0	2
SPAN		0	0	2	1.5	0	0	0	1	N/A	0	2
MFR, GEDIR		0	0	2	1.5	0	0	0	0	N/A	0	0
GOAFR		0	0	4	1.5		0	0	0	N/A	0	0
SAR	QoS	1	0	2	3	1	0	1	1	N/A	0	0
SPEED		1	0	2	3	1	0	0	0	N/A	0	0

Note: 1.5: Low, 2: Limited, 3: Moderate, 4: High

- If the estimation matches, then the messages will be updated in the system.
- The message updated in the system is known as resent; if this is not happening, then the process will not move forward.

The steps mentioned above will occur to some nodes with multiple updates; this is due to nodes nearer to BS will be updated more times than those located in distant from the BS. This process will be avoided in the MCFA with the help of the back-off algorithm.

3.6 *Gradient-Based Routing (GBR)*

GBR protocol is known to be suitable for the networking system, where each node maintains its gradient representing the direction towards a neighbourhood node to reach a sink. The nodes will remember the height of the node, which is known as the minimum number of hops needed to reach the base station. The difference of height between the nodes and neighbourhood node will be marked as gradient in the link. A packet with larger gradient will be passed to the link. GBR utilizes various processes such as aggregation of data and spreads the traffic eventually to the system in order to avoid traffic in the networking system. Multiple nodes passing the node is known as the relay node. This relay node will combine data on the basis of function. Gradient-based routing uses three different data dissemination methods: a Stochastic, energy-based and stream-based scheme.

3.7 *Information-Driven Sensor Querying and Constrained Anisotropic Diffusion Routing*

[7] proposed a framework utilizing 2 routing protocol known as the IDSQ (information-driven sensor querying) & CADR (constrained anisotropic diffusion routing)[7]. Constrained anisotropic diffusion routing is the general form of DD. The main concept of CADR is to query the data in the system and to route the data in the network for the process of gaining more information and reducing the bandwidth.

3.7.1 Cougar

It is another DC protocol. The energy consumption in COUGAR is done through data aggregation. In COUGAR, a social type of architecture is incorporated in the node system, where the leader node is selected to perform data aggregation and transmit the data to the base station. The base station segregates the queries to the nodes, and necessary steps must be taken for those queries as well as newqueries to be computed and sent to the relevant node. The planning structure of the queries may also help in selecting a leader for that particular query. This protocol helps to work with efficient energy for computation and in huge data generating process. Independent data queries method is provided in the COUGAR system.

3.7.2 Acquire

In ACQUIRE, multiple nodes will work when complex queries are required to perform by the system. In DD, multiple query handling techniques was not incorporated because of energy consumption level, and it works on the principle of flooding-based query system for aggregating queries. Responses will be given efficiently for queries in this method with the usage of lookahead (d) parameter. With equal lookahead, ACQUIRE will behave similar to that of flooding query mechanism. The queries need to travel a long distance if the size of d is small.

A mathematical model was needed to find the optimal value of lookahead in the system for nodes to get more neighbourhood nodes. The queries in this process will be forwarded based on an efficient node that may satisfy the needs of query or sometimes it may be picked randomly. Selection of next node will be based on information gained using IDSQ or CADR or through the information from the rumour routing protocol.

3.7.3 Energy-Aware Routing

Energy-aware routing protocol is energy-saving protocol that is utilized for minimizing the energy of the overall system consumed by wireless network sensors. This protocol is further classified as energy savers or energy balancers. This protocol initiates to know the connection between all nodes and source and the cost required for handling the queries, with the obtained information; later it will be updated in the table. The path that requires high cost is removed from this routing table, and the present table is forwarded with neighbourhood nodes that are proportional to the current cost. The forwarded table will send data to reach the nodes with a cost proportional to that of the node cost. Local flooding must be performed utilizing destination nodes in order to keep the path alive. When comparing the present protocol to DD, it shows an energy consumption level of 21%, and it also increases the life of the network system to 43%. This protocol needs to gather information about the destination, and address set-up of the destination must be done which involves complexity in setting up route for the network when compared to DD.

3.7.4 Routing Protocols with Random Walks

Random walk-based routing protocol is a probabilistic protocol in which each node selects randomly the neighbourhood nodes to forward the data packet. The path formed through this process is known as random walk (RW). The nodes can be turned on/off at multiple number of times randomly. The nodes can be identified with a unique process without the need of information. The arranging pattern of the nodes are so perfect such that the nodes will fall on the crossing point on a grid on a plane but topological order may be irregular. The BELLMAN-FORD algorithm is capable of computing shortest distance from the source to all the other nodes. The node nearer to the neighbourhood node will be chosen as an intermediate node. This protocol is simple but little more information about the nodes need to be maintained. Different routes are chosen for nodes with similar source to destination. One main factor of this protocol is that the topology of the network will not be practical.

3.7.5 Hierarchical Routing (HR)

Hierarchical routing protocol was first proposed for the system of wireless sensor network which are efficient structure of scalability and communication system. This system has efficient energy consumption. This system uses nodes with lesser energy for sensing for achieving the target. This process tries to assign work for CH (cluster head) by contributing to the entire networking system's life, energy, and scalability. HR is the significant method for reducing energy consumption within the cluster network, ability to perform data aggregation, and ability to decrease message transmission process in the base station.

HR is a two-layer routing system:

- The first layer may be used for selecting the CH.
- The second layer utilized for selecting the route.

The main concept of this system is not always about routing; it is about to whom/when/where to send the processed information, allocating channels for broadcasting, etc.

3.7.6 LEACH Protocol (Low-Energy Adaptive Clustering Hierarchy)

[1] proposed a framework utilizing HR for sensor network called as the LEACH. It is a clustering based protocol where distributive clustering formation is also included in the system. It selects the nodes as CH randomly and the role of CH will be rotated in order to reduce energy loading among the sensors in the networking system. In this LEACH protocol, the nodes arriving from other data node and the CH nodes altogether send packets to the base station by which the number of messages transmitted to the base station will be reduced. LEACH utilizes Time-division Multiple Access (TDMA) for CDMA/Media Access Control (MAC) for the purpose of reducing collisions. The collection of data in this protocol happens on the basis of centralized system and it happenson a timely basis. This system will be monitored with the help of a centralized system.

LEACH along with negotiation was proposed by [1]. The aim for this protocol is to proceed transferring of data with higher negotiation with the help of metadata in SPIN. The data with new information is only transmitted to the cluster head beforehand of transmitting to the base station.

3.7.7 Power-Efficient Gathering in Sensor Information Systems

[8] proposed a framework utilizing LEACH with a new enhancement in it which is known as PEGASIS (Power-Efficient Gathering in Sensor Information Systems); it is almost similar to optimal chain protocol. The idea behind enhancement of this new protocol is trying to enhance the life of the networking system, and communication between the neighbourhood nodes and base station by the other nodes must occur without any hindrance. The nodes communicating with base station ends and the process repeats as well. This protocol provides a system that is efficient in using energy for transmitting data in the base station. The two important objectives of PEGASIS are:

- The lifetime of the nodes is increased with the help of collaborative technique.
- The communication between the nodes that are closer occurs in order to reduce the bandwidth, which are required for communicating with the other nodes in the system.

In this present protocol, the clustering formation is avoided, and one node is needed in the chain to transmit data to the base station without the usage of multiple node.

The neighbourhood nodes in the system are identified by using signal so that the nodes that are closer will be identified easily and the signal strength will be adjusted to make it audible to other nodes.

The construction of chain in this protocol is done using greedy fashion. The outcomes from this protocol shows that the lifetime of the network will be increased twice the time when compared to LEACH. The further modified PEGASIS is known as the Hierarchical-PEGASIS.

3.7.8 Threshold-Sensitive Energy-Efficient Protocols

[9] proposed a framework utilizing TEEN (Threshold-Sensitive Energy-Efficient Sensor Network Protocol) & APTEEN (Adaptive Periodic TEEN). This is developed for use during critical time of the application. The transmission in TEEN is done regularly but transmission of data is done irregularly. The cluster head sends its CM to the harder threshold with the value of sensed attribute, the value of the softer threshold is with small change.

In harder threshold, the transmission will be lesser because only the attribute with interest range will be involved. In softer threshold, the transmission occurs with very lesser number because of the change in the value of attribute.

The main disadvantage of this system is that if the threshold is not obtained, then the communication between the nodes will not occur and the data from the network will not be transmitted to the user.

3.7.9 Small Minimum Energy Communication Network (MECN)

[10] proposed a framework utilizing MECN which is used for identifying the sensor with low-power global positioning system. The relay region of the nodes is identified using MECN. The relay region is the one with details of the neighbourhood node, which may help in transmitting to the node that helps save energy.

The significant idea of MECN is to identify subnetwork with less nodes and also the one which utilizes less power during the process of transmission between the two nodes. MECN can adapt to the failures which includes new sensor to the networking system. A small MECN is known as SMECN. In this any node can transmit data to other node but this process is not possible all the time. The subnetwork developed with the help of SMECN utilizes minimum energy while relaying than that of the one constructed by MECN.

3.7.10 Self-organizing Protocol (SOP)

[1] proposed a framework utilizing the self-organizing protocol. The sensors of this protocol are made of momentum. Some sensors will investigate the surrounding and then data will be forwarded to the destination node which acts as routers. The router node is always in momentum which is considered as the strong support system for communication. The data collected from the nodes are then forwarded with the help of router to the nearby base station node. All the nodes in the system must be capable of reaching the router system so that it can be a part of the network. A routing model for addressing the nodes has been developed. These nodes are identified using the address of the router node, which is connected to the node.

3.7.11 Sensor Aggregates Routing

[11] proposed a framework utilizing sensor aggregate routing. The main aim of this protocol is to manage the environment and the activity needed to be attained by this system. The sensor combines nodes in the system for grouping those nodes to perform certain collective task. The formation of these nodes are based on collecting for the purpose of sensing as well as for communication. The sensors in this field are separated as clusters based on their sensing capacity and strength. The cluster leader for the localized nodes is selected.

The election for selecting the leader may occur only if the information is exchanged between the neighbourhood sensors. The leader-based algorithms have an assumption that the unique leader may know the region of collaboration. An important algorithm that has been proposed is known as Distributed Aggregate Management (DAM). This is responsible for the formation of sensor aggregate for the purpose of monitoring the targeted task. This protocol consists of decision predictor for all nodes to decide whether the nodes participate in the aggregate function or not for exchanging message.

The second protocol developed in this section is the EBAM known as the Energy-Based Activity Monitoring. The nodes' energy level will be measured using the signal at the specific area. The third algorithm, Expectation-Maximization Like Activity Monitoring (EMLAM), eliminates the constant and equal target energy-level assumption. EMLAM assesses the signal energy and target positions using received signals and utilizes the resultant estimates to calculate how signals from the targets may be mixed at each sensor.

4 Routing Protocols Based on Protocol Operation

The following section tries to highlight more routing protocols and their functions, and these may be under the category of routing principle.

4.1 Multipath Routing Protocols

This section describes the protocols with multipath, and enhancement in the network performance has been noted. The fault tolerance of these protocols must be measured so that alternative path can be chosen for the node to reach from the source to the destination if any mistake arises in the regular main path [12].

By using multipath to reach from source to destination requires more expense for communication as well as involves more energy, and higher level of traffic is generated in the networking system. The alternation route must be kept alive so in order to do so messages are send at a regular basis. The reliability of the network is increased with higher expense for maintaining the overhead cost of the alternative route.

[13] proposed a framework utilizing routing model that routes the data for the nodes to have high residual energy. But this path is not fixed because new better path will be taken if needed. Till then, the main path will be used; if the energy falls below the energy of the alternative path, then that will be taken immediately. Through this method, the nodes utilizing the main path will not exhaust so easily through continuous usage of the same path again and again so that life of the path will be longer than the expected one.

[14] proposed a framework utilizing sub-optimal path in order to increase the life of the networking system. This path is chosen based on the energy consumption by the path; if it consumes lesser energy, then it may be selected. The path that consumes more residual energy while routing data in the network may also utilize

more energy as well, so trade-off occurs in this situation to minimize the power consumption and residual energy consumed by the network.

[11] proposed a framework where the residual energy is less a little by which more paths which are highly energy efficient will be selected.

[15] proposed a framework utilizing multiple paths in order to enhance the reliability of Wireless Sensor Network. Even data will be delivered in an unreliable surrounding. The reliability of the network can be increased by having multiple paths from the source to the destination and also for sending similar packets in the path. But traffic will increase simultaneously.

The trade-off occurs here also between the level of traffic and the network reliability. The trade-off is determined utilizing redundancy which is independent in the multiple path. The main idea behind this is to divide the packets into multiple sub-packets, and the sub-packets will be sent through any one of the multiple [3]. The advantage of this system is that if any one of the sub-packet is lost, the messages can be reconstructed. It is found that the one with highest failure in the nodes utilizing multiple path than the optimal value will try to maximize the total probability of failure. DD is a capable one for robust multiple path routing and delivery [5].

4.2 *Query-Based Routing*

In this routing, the nodes at the destination will generate a query for the purpose of data through the node for the network, and the nodes with perfect data matches itself with the perfect query of the node that was responsible for initiating the query first. These queries are mentioned with the use of simplistic language or higher end query language. DD is the perfect example for this type of routing. In DD, the nodes in the base station send out interested message to the sensor [5].

Rumour routing process utilizes a set of agents that have long life in order to create path, which are used to direct towards the event they come across. When the nodes come across the path that leads to the vent which has never met before, then at that time a new route will be processed which is leading to that event. The nodes will try to maintain list in order to know the neighbourhood node, and the events will be marked in the table. The nodes can circulate an agent; this agent contains a relay list containing the events table which will be synced with other node [16]. These agents have a certain lifetime in the networking system; after that period, they will die. The nodes will not circulate any query to the agents until the correct route is reached to the agent. Without proper routes, the nodes will transmit the queries in random direction. The node will wait till the query to reach the perfect destination.

4.3 Negotiation-Based Routing Protocols

[3] proposed a framework called the SPIN (Sensor Protocols for Information via Negotiation). These protocols come in different varieties and are designed for addressing the deficiency of flooding and gossiping. They are highly different when compared to the conventional data-centric protocol. In this protocol, the user can query any nodes in the system and get information from them. This protocol uses the data from the nodes those are nearer to it and those with similar data, so only data distribution will be the further process in it. The Sensor Protocols for Information via Negotiation uses the current energy and run the network and the working processes are done with energy remaining in the system.

The working structure of this protocol is based on time-driven pattern through which the information is distributed to the entire network, even if the information is not needed for the users. The main aim of SPIN protocol is to address the deficiency in classic flooding through the process of negotiation and adaption of resource.

4.4 QoS (Quality of Service)-Based Routing

In the Quality of Service routing protocol, there must be a balance between consumption of energy level as well as data quality. The network must at least satisfy some of the factors of the QoS before delivering the data to the base station.

[14] proposed a framework utilizing the SAR known as the Sequential Assignment Routing. This is the first routing protocol of Wireless Sensor Network to introduce the Quality of Service to the decisions in routing.

The decision of routing in Sequential Assignment Routing is based on three dependent factors:

- Energy resource
- Quality of service in each route path
- The preference level for each packet

For single routing failure, multiple routes and local path system are utilized in this process.

The Sequential Assignment Routing calculates the Quality of Service metrics as the source of additional Quality of Service metric and the weight of coefficient involved in the higher preference in the packet. The main aim of Sequential Assignment Routing is to reduce the weightage of Quality of Service throughout the entire lifetime of the networking system. Recomputation is required if the node fails due to the change in the topology of the protocol.

A handshake method on the basis of localized path restoration between the neighbourhood node is utilized to recover the nodes from failure. The failure is retrieved using enforcing upstream and downstream in the nodes in the routing table. From the result of Sequential Assignment Routing, it shows that lesser power was consumed than that of the energy of the metric algorithm. Sequential Assignment Routing utilizes multiple path from the node to reach the base station. This protocol has good fault-tolerant capacity and recovery from failure as well. One of the disadvantage of this system is that it suffers with overhead issue when more number of nodes are involved.

4.5 Coherent and Non-coherent Processing

Processing data for the operation of WSN is very important. Different methods of data processing are done through various routing techniques [17]. The nodes will work collectively with the other nodes in the network while data processing. The best example of data processing methods in Wireless Sensor Network are framed as coherent and non-coherent data processing. In the second method of data processing, the nodes will first process the fresh data before sending it to the other nodes for next step. The nodes responsible for performing the next step is known as the aggregator. In the first data processing method, the data will be passed to the aggregator after low level of processing. In order to perform task with minimum energy, coherent data processing is usually picked.

5 Conclusion

The routing process in Wireless Sensor Network is typically the new research concept which is highly preferred nowadays. In this present research, several routing techniques have been explained with reference to the WSN which is capable for having a network with prolonged lifetime. The routing protocols in this research have been classified into three categories as Hierarchical, Flat and Location-based routing protocol. The above routing protocol was further divided into next level of category based on the working structure of the protocol. Several wider concepts have also been discussed in this chapter including the trade-off and the merits and demerits of the above routing protocol. From this research, we found that many other areas in routing protocol must be addressed in the WSN-based networking system. The further level of research area in this present chapter has also been highlighted here.

References

1. Pantazis, N. A., Nikolidakis, S. A., & Vergados, D. D. (2012). Energy-efficient routing protocols in wireless sensor networks: A survey. *IEEE Communication Surveys and Tutorials, 15*(2), 551–591.

2. Kocherla, R., & Vatambeti, R. (2020). Review and analysis of energy-efficient routing in wireless sensor networks. *Journal of Computational and Theoretical Nanoscience, 17*(7), 2969–2974.
3. Warrier, M. M., & Kumar, A. (2016). Energy efficient routing in wireless sensor networks: A survey. In *2016 International Conference on Wireless Communications, Signal Processing and Networking (WiSPNET), Chennai* (pp. 1987–1992). https://doi.org/10.1109/WiSPNET.2016.7566490.
4. Kumar, A., et al. (2017). Location-based routing protocols for wireless sensor networks: A survey. *Wireless Sensor Network, 9*(1), 25–72.
5. Ogundile, O. O., & Alfa, A. S. (2017). A survey on an energy-efficient and energy-balanced routing protocol for wireless sensor networks. *Sensors, 17*(5), 1084.
6. Goyal, D., & Tripathy, M. R. (2012). Routing protocols in wireless sensor networks: A survey. In *2012 Second International Conference on Advanced Computing & Communication Technologies*. IEEE.
7. Ghaffari, A. (2014). An energy efficient routing protocol for wireless sensor networks using A-star algorithm. *Journal of Applied Research and Technology, 12*(4), 815–822.
8. Chen, J., et al. (2017). Wireless sensor networks based on modular arithmetic. *Radioelectronics and Communications Systems, 60*(5), 215–224.
9. Pour, N. K. (2016). Energy efficiency in wireless sensor networks. arXiv preprint-Cornell University Library. *arXiv* preprint arXiv:1605.02393, 1–158.
10. Tabibi, S., & Ghaffari, A. (2019). Energy-efficient routing mechanism for mobile sink in wireless sensor networks using particle swarm optimization algorithm. *Wireless Personal Communications, 104*(1), 199–216.
11. Darabkh, K. A., et al. (2017). A novel clustering protocol for wireless sensor networks. In *2017 International Conference on Wireless Communications, Signal Processing and Networking (WiSPNET)*. IEEE.
12. Asif, M., et al. (2017). Quality of service of routing protocols in wireless sensor networks: A review. *IEEE Access, 5*, 1846–1871.
13. Faheem, M., et al. (2015). EDHRP: Energy efficient event driven hybrid routing protocol for densely deployed wireless sensor networks. *Journal of Network and Computer Applications, 58*, 309–326.
14. Kumar, N., & Singh, Y. (2017). Routing protocols in wireless sensor networks. In *Handbook of research on advanced wireless sensor network applications, protocols, and architectures* (pp. 86–128). IGI Global.
15. Rawat, P., et al. (2014). Wireless sensor networks: A survey on recent developments and potential synergies. *The Journal of Supercomputing, 68*(1), 1–48.
16. Samara, G., & Aljaidi, M. (2019). Efficient energy, cost reduction, and QoS based routing protocol for wireless sensor networks. arXiv preprint arXiv: 1903.09636.
17. Kumar, N., Singh, Y., & Singh, P. K. (2017). Reputation-based energy efficient opportunistic routing for wireless sensor network. *Journal of Telecommunication, Electronic and Computer Engineering (JTEC), 9*(3–6), 29–33.
18. Pantazis, N. A., Nikolidakis, S. A., & Vergados, D. D. (2013). Energy-efficient routing protocols in wireless sensor networks: A survey. *IEEE Communication Surveys and Tutorials, 15*(2), 551–591., Second Quarter. https://doi.org/10.1109/SURV.2012.062612.00084.
19. Darabkh, K. A., & El-Yabroudi, M. Z. (2017). A reliable relaying protocol in wireless sensor networks. In *2017 European Conference on Electrical Engineering and Computer Science (EECS)*. IEEE.

Emperor Penguin Optimization Algorithm and M-Tree-Based Multi-Constraint Multicast Ad Hoc On-Demand Distance Vector Routing Protocol for MANETs

M. Deva Priya, M. Rajkumar, S. Karthik, A. Christy Jeba Malar, R. Kanmani, G. Sandhya, and P. Anitha Rajakumari

Abstract Multicast based routing in ad hoc networks is considered essential for attaining reliable data dissemination. However, trusted data transmission can be achieved by using optimal multicast trees that aid in better performance of the network. Further, prolonging network lifetime is yet another issue that needs to be concentrated for sustained connectivity. In this chapter, Emperor Penguin Optimization Algorithm and M-Tree-based Multicast Ad hoc On-demand Distance Vector Routing (EPOA-MT-MAODV) protocol is proposed for optimal selection of multicast routes for enhancing the lifetime of the network. This proposed EPOA-MT-MAODV protocol utilizes the merits of exploitation and exploration inherited from Emperor Penguin Optimization Algorithm (EPOA) with the estimation of multifactor, path inclusion and destination. It focuses on delay, minimum distance, link stability and energy for optimal selection of optimal tree. The simulation results of the proposed EPOA-MT-MAODV protocol confirm better performance in terms of energy consumption and Link Lifespan Time (LLT) for varying number of mobile nodes and mobility speeds.

M. Deva Priya · G. Sandhya
Department of Computer Science and Engineering, Sri Krishna College of Technology, Coimbatore, Tamil Nadu, India
e-mail: m.devapriya@skct.edu.in; sandhya.g@skct.edu.in

M. Rajkumar · A. Christy Jeba Malar (✉) · R. Kanmani
Department of Information Technology, Sri Krishna College of Engineering and Technology, Coimbatore, Tamil Nadu, India
e-mail: rajkumarm@skcet.ac.in; a.christyjebamalar@skct.edu.in; r.kanmani@skct.edu.in

S. Karthik · P. Anitha Rajakumari
Department of Computer Science and Engineering, SRM Institute of Science and Technology, Ghaziabad, Uttar Pradesh, India
e-mail: karthics1@srmist.edu.in; anitharp@srmist.edu.in

A. Haldorai et al. (eds.), *3rd EAI International Conference on Big Data Innovation for Sustainable Cognitive Computing*, EAI/Springer Innovations in Communication and Computing, https://doi.org/10.1007/978-3-030-78750-9_7

Keywords Multicast tree · Emperor Penguin Optimization Algorithm · Multicast routing · Link lifespan · Minimum distance · Path-enclosure factor · Terminal flag

1 Introduction

Mobile ad hoc networks (MANETs) do not demand any infrastructure and are self-organized wireless networks without any centralized control. The multi-hop wireless links permit the nodes in arbitrary motion to interchange data among themselves. Communication between the nodes can be built through intermediate nodes even if they are beyond the communication range. Though the nodes are capable of building a network with their self-establishing capacity, these nodes face challenges in terms of resources, transmission power and battery life [1]. It is likely for the nodes to function as intermediary nodes or hosts. It involves an autonomous multi-hop network with mobile nodes that can be used in many real-time applications. The important challenge faced in a MANET is finding paths with improved Quality of Service (QoS). Limited resources may lead to instability in the paths connecting the source and the destination, making communication tedious. Ensuring energy efficacy is a challenging factor in the effective positioning of MANETs as nodes are anticipated to rely on moveable and restricted energy sources [2]. Furthermore, energy preservation is exceptionally challenging in multi-hop scenarios, where the nodes involve energy in routing packets and ensuring network connectivity [3, 4].

Multicasting supports transmission of datagrams to a collection of mobile nodes recognized by a single multicast address, and hence is envisioned for group-based computing. Multicast supports one-to-many and many-to-many communication, which preserves bandwidth by sending data to numerous receivers wherein packets are replicated only at Cluster Heads (CHs). Multicasting enables grouping of receiver nodes and optimizing bandwidth use. Multicast communication in MANETs is challenging than offering services in the fixed Internet, as the services intended for wired networks may not be appropriate for MANETs. Multicasting is useful in applications involving dissemination of newsletters, software and stock quotes, audio/video conferencing, online education, online games and communications among military troops or rescue teams. Multicast routing reduces bandwidth consumption in links, sender and receiver processing and delivery delay in contrast to broadcasting, wherein a single sender forwards datagrams to numerous hosts. Multicast routing supports multiple senders and receivers. They play a predominant role in emergency communications where the network is built temporarily and rapidly. As the nodes move arbitrarily, the routing protocols must be extremely efficient and dependable to ensure packet delivery. Depending on the method of data delivery, the prevailing multicast routing protocols are categorized into tree and mesh-based protocols.

2 Related Work

In this section, the works done by various authors related to multicast routing protocols for MANETs are discussed.

Yen et al (2008) [5] have focused on multicast routing with numerous QoS constraints in MANETs. It is an NP-complete problem involving many QoS constraints. The authors have designed an energy-proficient Genetic Algorithm (GA)-based scheme to deal with the constraints. A routing algorithm based on the source tree is designed, and a multicast tree involving shortest distance is built to lessen the delay by involving a small population in GA. Some nodes are involved in finding routes. The genetic sequence and topology encoding are enhanced to extend the lifespan of nodes that compute the Residual Energy (RE) of the nodes in the multicast tree. Clustering enables reduction of the number of mobile nodes involved in multicasting, which leads to decrease in control overhead. Astier et al (2009) [6] have propounded Restful Stability-based Insomnious Distributed Sensors (RSIDS), which takes the robustness and RE of nodes at proximity when choosing prominent nodes including CHs and Gateways (GWs). It involves passive clustering, and nodes with high RE are elected as the prominent nodes, thus overcoming the need for re-clustering. This reduces the overhead and packet loss, and improves the network lifetime.

Hosts with many neighbours are selected as Backbone Hosts (BHs) to send packets. The BHs act as bottlenecks in the network. As the mobility of hosts is not considered for BH selection, these techniques involve more packet losses if extreme mobile hosts are appointed as BHs. In the propounded two-tier protocol, hosts with scarcer hops and lengthier residual connection time are chosen as BHs. The main aim is to construct small and steady multicast routes with a robust two-tier infrastructure involving less packet losses. Some schemes are implemented as probing schemes, but they are incapable due to increased overhead and less response. Some schemes that consider routing and link details to decrease the inadequacy are also open to bandwidth damage problems. Hu (2010) [7] have constructed bandwidth content multicast trees for applications in MANETs. None of the existing routing and multicast protocols in MANETs use two-tier infrastructure to circumvent inadequacy of flooding. This scheme involves reduced number of forwarders. Jabbehdari et al (2010) [8] have propounded a multicast routing protocol that adds two features with QoS considerations. The primary feature is that if a node lacks bandwidth, it can be recovered by gathering the feedback from the application. The secondary feature is the capability of changing the value of the timer and delay based on the mobility speed.

A cluster-based multi-source multicast routing protocol is proposed by Hwang et al (2011) [9] to support effective multicasting in a multicast environment with multiple sources. It involves formation of common forwarding clusters and routing to conserve advancing efficacy and robustness. The propounded model offers better Packet Delivery Ratio (PDR) for a multicast environment with multiple sources. A weight-based clustering scheme is proposed to select appropriate nodes as CHs to monitor nodes and members in the multicast group. A multicast route is built by arranging transmitting clusters. An upkeep mechanism to adjust to topology modifications instigated by mobility of nodes is also proposed. Santhi & Nachiappan (2011) [10] have presented a multicast routing mechanism that finds a path based on fuzzy cost by involving QoS metrics like delay, bandwidth and number of nodes. Biradar & Manvi (2011) [11] have applied a collection of static and mobile agents for multicast routing to develop a backbone as a steadfast ring and identify multicast routes. It applies convex hull method to find the network boundary that assists in forming a consistent ring. Kharraz et al (2012) [12] have presented a dynamic multicast routing protocol with proficient route discovery that enhances the performance and efficiency of On-Demand Multicast Routing Protocol (ODMRP). The scheme supports limited flooding by enhancing multicasting based on the delay of nodes. The contributing nodes are formed using M/M/1 queueing methods. Wang (2012) [13] has propounded a Power-aware Dual-Tree-based Multicast Routing Protocol (PDTMRP) wherein the nodes are arbitrarily categorized into two groups, namely, group-0 and group-1. To attain load balancing, two multicast trees are built. Kim et al (2013) [14] have dealt with the construction of Steiner trees to support multicast routing. The mechanism produces a weighted multicast factor by proficiently merging independent features like cost and delay. The proposed Weighted Parameter for Multicast Trees (WPMT) includes weights {0, 1}. If the weight is '0', the delay of the multicast tree may be comparatively lesser than that of the trees obtained as the weight moves towards '1'. When the weight lines to 1, the cost of the tree is reduced in contrast to other trees. Li et al (2014) [15] state that though the tree-based protocols support better forwarding by consuming low bandwidth, they are less robust. The influence of the load of a network on MAODV protocol is focused and Multicast Ad hoc On-demand Vector with Backup Branches (MAODV-BB) which enhances the robustness of the protocol by merging merits of the tree and mesh structure is proposed. It is capable of updating shorter branches and also building a tree with backup branches. This model is capable of dealing with heavy loads.

Kumar & Sachin (2016) [16] have dealt with multicast routing to focus on the challenges like dynamic nature of topology, bandwidth inadequacy and energy limitation leading to reduced lifespan. An energy-proficient, scalable and resilient protocol is highly entertained. A trusted link should be established between nodes in group communication. The authors have propounded a mesh-based multicast routing protocol using consistent neighbour nodes. Only dependable nodes are included for routing, and the nodes with reduced reliability pair factor computed based on the RE of nodes are ignored. Multicasting involves various factors including vibrant nature, bandwidth scarcity and reduced lifetime of devices. Yadav & Tripathi (2017) [17] have propounded Quality of Service (QoS)-based Multicast Routing Protocol called Reliable Neighbour Nodes Selection Scheme (QMRPRNS) which offers consistent neighbour node selection.

Tavizi & Ghaffari (2018) [18] have propounded a Reliable and Energy-Aware Multicast Ad hoc On-demand Distance Vector (REA-MAODV) routing protocol. It supports better consumption and smaller tree branches, and a multicast tree with associated branches. It includes the following stages: Choosing a path with better energy consumption, choosing process and incorporating branches for support and upholding scheme of multicast tree. Hassan et al (2019) [19] have employed African Buffalo Optimization (ABO) to enhance the QoS of the AODV routing protocol. Optimal paths are selected in the AODV routing protocol. Incorporation of ABO reduces the delay and the energy involved in routing. ABO is proficient in imitating and using effectual organization and communication method of the herd during relocation. During decision-making, they establish the voting characteristic, and this decision manages the mobility of nodes. The sounds 'maaa' and 'waaa' are used in their movement for exploring as well as exploiting. The buffaloes remain in the same location as they have sufficient pasture and protection. Instead, the 'waaa' sound is used for exploring new locations as the pasture in the present location may be insufficient. The search of buffaloes is improved by making use of these sounds in order to get access to fruitful regions. The nodes involved in transmission face energy constraints, and lifetime becomes a predominant factor. Further, the routes should own maximum RE and involve less energy for transmitting data. Papanna et al (2019) [20] have propounded an Energy Efficient Lifetime Aware Multicast (EELAM) route selection mechanism designed using the dynamic Genetic Algorithm (GA). It functions based on tree topology that adopts evolutionary computation mechanism which plays a predominant role in choosing optimal intermediary nodes with increased RE. The fitness function designed for dynamic Genetic Algorithm (GA) improves the energy consumption ratio, RE and multicasting possibility.

3 Emperor Penguin Optimization Algorithm and M-Tree-Based Multi-Constraint Multicast Ad Hoc On-Demand Distance Vector (EPOA-MT-MAODV) Routing Protocol for MANETs

In this section, Emperor Penguin Optimization Algorithm and M-Tree-based multi-constraint Multicast Ad hoc On-demand Distance Vector (EPOA-MT-MAODV) routing protocol are proposed for multicast routing. It includes two stages: (i) M-tree construction using DIVisional-based Clustering (DIVC) and (ii) Multicast routing using Emperor Penguin Optimization Algorithm (EPOA). In DIVC, an M-Tree is constructed using Route Request (RQ) and Route Reply (RP) scheme by using three limitations namely, multi-feature, path-enclosure factor and terminal flag. Once the M-tree is constructed, the EPOA chooses optimal paths from the available paths in the M-tree. The main goal of the EPOA algorithm is to select the solution with the best fit as the optimal path for communication. The fitness function is based on the energy, delay, Link Lifespan Time (LLT) and distance. The block diagram of the proposed system is shown in Fig. 1.

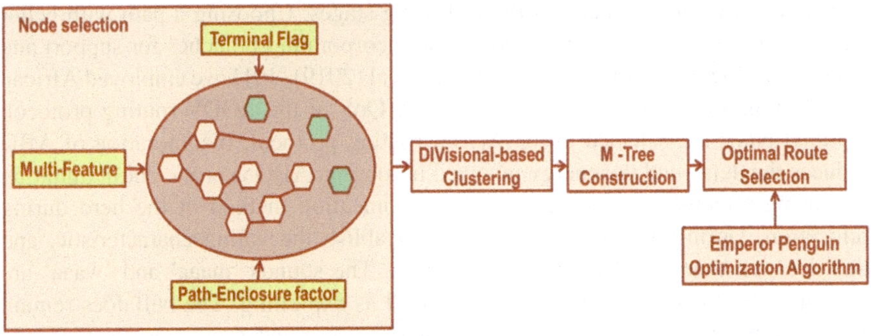

Fig. 1 Block diagram of the proposed EPOA-MT-MAODV protocol

3.1 DIVC-Based M-Tree Formation

The DIVC clustering scheme is a divisive scheme that involves top-down method. The data is grouped into clusters, further divided into clusters until the entire set of data is grouped into distinct clusters. The clusters are divided based on various standards or limitations which are satisfied by a node in the cluster. The algorithm is based on the following constraints namely, multi-feature, path-enclosure factor and terminal flag to perform clustering and build the M-tree.

The source that wants to forward a packet sends a RQ packet to its adjacent nodes in the communication range. Every node verifies whether the constraints are fulfilled to advance the data. The nodes that fulfil the conditions become a part of communication.

Multi-Feature The multifactor is based on different factors including energy, LLT and distance between two nodes in a cluster.

$$\text{Multi} - \text{Feature} = \left[E_{\int \text{node}} + \frac{LLT}{LLT_{\text{NORM}}} + \left[1 - \frac{D_r}{1 - DT_{\text{NORM}}} \right] \right] * \frac{1}{3} \tag{1}$$

where

$E_{\int \text{node}}$ – Energy of a node
DT – Distance between nodes
LLT – Lifetime of a link
LLT_{NORM}, DT_{NORM} – Normalized LLT and DT

Terminal Flag It reports the destination during communication to choose a path. During path discovery, the scheme chooses nodes by examining whether it is the destination by using a flag. If it is a destination, the terminal flag is set and a path is created. When flag is zero, the node forwards RQ until the destination is reached.

Path-Enclosure Factor It finds the node in a path that is involved in communication once. Repeating paths can be circumvented, thereby reducing the time taken for

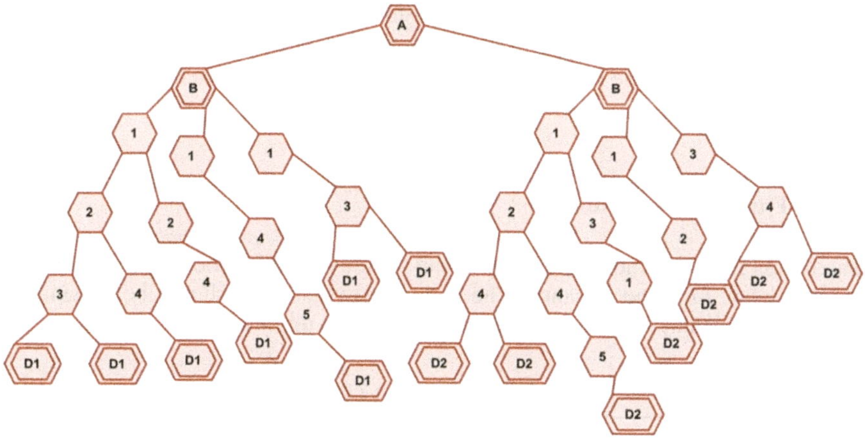

Fig. 2 Construction of M-tree

processing and dropping false information. It has a preset value to accept or reject nodes along the path.

Only if the constraints are satisfied, a node forwards RQ packet to the destination nodes through the intermediary nodes. Once the RQ packet is received, the destination commences the RP stage. The destination sends the RP packet to the source using the inverse path data in the RP packet. The in-between nodes forward it until the source receives the RP packet. Thus, the M-tree is built with the routes formed between the source and the destination. Figure 2 shows the M-tree built using DIVC as an illustration.

3.2 *EPOA-Based Multicast Routing*

The optimum choice of multicast routing to forward data packets is discussed below. In the proposed scheme, a multicasting routing scheme is formulated based on EPOA algorithm that focuses on enhancing the process of searching. The EPOA algorithm is an optimization method developed based on the inspiring features of emperor penguins. The fitness assessment of EPOA algorithm concentrates on the construction of an optimal multicast tree for transmission-based fitness assessment and multicast routing.

3.2.1 Solution Encoding

Solution encoding is the easiest method of demonstrating the algorithmic process. The goal of multicast-based routing algorithm is to send the data packets from the source to the anticipated number of destinations. Let 'D_1' and 'D_2' represent the destinations which accept information from the source 'S'. Many paths exist from the source, and the one with the best fitness values is chosen as the optimum solution.

3.2.2 Fitness Formulation

The fitness function finds the quality of the solution. In this problem, it is based on energy, delay, *LLT* and distance. The node energy and *LLT* should be maximum, while the delay and distance among nodes should be minimum.

$$F = \left(\frac{f_1}{2N}\right) + \left\{\frac{1}{3 * f_1 * 2}\left[\sum_{i=1}^{D}\sum_{j=1}^{p_i}\sum_{n=1}^{N_j}\left(E + \frac{LLT_{(n,n+1)}}{LLT_{\text{NORM}}} + \left[1 - \frac{DT_{(n,n+1)}}{DT_{\text{NORM}}}\right]\right)\right]\right\} \tag{2}$$

where

N – Number of nodes in the network
D – Number of destinations
P – Number of paths to reach the destination
E – Energy of a node
N_j – Number of nodes in the jth path
f_1 – Number of nodes along a path given as

$$f_1 = \sum_{i=1}^{D}\sum_{j=1}^{p_i}\sum_{t=1}^{N_j}|N_j| \tag{3}$$

An efficient algorithm demands improved node energy so that the node becomes a part of the communication. The mobility of a node and dynamic topology leads to energy depletion. It is mandatory to preserve the energy in a node throughout the communication to the maximum extent possible. Further, the link should be alive for a longer duration. *LLT* impacts the lifespan of a path to increase the lifetime of a network. The node's mobility affects the lifespan of the links. It can be compared to the energy in the nodes which necessitates it to be maximum to improve the *LLT*. In addition, the Distance Metric (*DT*) between the nodes should be retained to a minimum. In the mobility model, the distance between the nodes is measured. The delay is directly comparable to the number of nodes in a route. It increases with the number of nodes.

4 Emperor Penguin Optimization Algorithm (EPOA)

Emperor Penguin Optimization Algorithm (EPOA) is a meta-heuristic algorithm, which is used in optimization scenarios based on their huddling behaviours that help them to survive in the Antarctic winter. This huddling behaviour includes four stages namely: (i) generation and estimation of the boundary of the huddle, (ii) computation of temperature profile existing around the boundaries of the huddle, (iii) estimation of the distance between the emperor penguins and (iv) effective mover relocation. Moreover, every penguin has equal probability to participate in the huddling behaviour according to the level of warmth. This EPOA completely focuses on the estimation of an effective mover. It is assumed that the huddle is located on the two dimensional L-shaped polygon plane. Initially, the EPOA randomly generates the boundary of the huddle. Then, the temperate profile associated with the huddle boundary is estimated. Further, the individual distances between the emperor penguins are determined in achieving the maximum degree of exploitation and exploration. Finally, the best optimal solution is attained (effective mover), and huddle boundary is recalculated such that the results are obtained to update the location of the search agents (penguins).

4.1 Generation and Estimation of the Boundary of the Huddle

The search agents (emperor penguins) are considered to traditionally position themselves during the huddling process over a grid boundary of polygonal shape. This huddling process assures that the search agent always has at least a minimum of two neighbouring search agents selected randomly from the complete huddle. In this context, the huddle boundary realized around the polygonal grid is determined based on the wind flow. But, the movement of penguins is always considered to be slower than the flow of wind. At this juncture, complex variables 'ψ'and 'δ' are used for presenting the randomly generated huddle boundary as shown in Eqs. (4) and (5).

$$\phi = \nabla\psi \tag{4}$$

$$F = \psi + i\delta \tag{5}$$

where 'F' corresponds to the polygon plane analytical function with 'i' representing the imaginary constant.

4.2 Computation of Temperature Profile Existing Around the Boundaries of the Huddle

In this phase, the emperor penguins construct the huddle in order to save energy and improve the ambient huddle temperature. This situation of huddle construction is achieved mathematically by assuming the initial temperature '$T_{\text{Huddle}} = 0$' with the polygon radius '$R_P > 1$' and temperature '$T_{\text{Huddle}} = 1$' with polygon radius '$R_P < 1$'. This profile of huddle temperature is capable of balancing the trade-off between exploitation and exploration based on different locations of the emperor penguins. In this context, the temperature profile identified around the huddle at temperature '$T_{\text{Huddle}} = 0$' is based on Eqs. (6) and (7).

$$T_{\text{Huddle}} = \begin{cases} 0, \text{if } R_P > 1 \\ 1, \text{if } R_P < 1 \end{cases} \tag{6}$$

$$T^{C}_{\text{Huddle}} = \left(T_{\text{Huddlebuild}} - \frac{\text{Iter}_{\text{Max}}}{\text{Iter}_{\text{Current}} - \text{Iter}_{\text{Max}}} \right) \tag{7}$$

where '$\text{Iter}_{\text{Current}}$' and '$\text{Iter}_{\text{Max}}$' represent the maximum number of iterations and current iterations, respectively. 'R_P' and '$T_{\text{Huddle\_build}}$' refer to the polygon radius and time incurred in determining the best optimal path in the network search space.

4.3 Estimation of the Distance Between the Emperor Penguins

In this step, the distance between individual emperor penguins and the superior optimal solution is estimated after the construction of the huddle boundary. The remaining emperor penguins or the other search agents are responsible for enhancing the positions using the existing best optimal solution determined mathematically based on Eq. (8).

$$\overrightarrow{D_{IEPQ}} = Abs\left(S_{F(EPQ)} \left(\overrightarrow{C_{PV(1)}} * S_{\text{Best}(EPQ)} - \overrightarrow{C_{PV(2)}} * S_{PVC(EPQ)} \left(\text{Iter}_{\text{Current}}\right)\right)\right) \tag{8}$$

where '$\overrightarrow{C_{PV(1)}}$' and '$\overrightarrow{C_{PV(2)}}$' are the collision prevention factors utilized for avoiding collision among the neighbourhood emperor penguins. '$S_{\text{Best}(EPQ)}$' and '$S_{PVC(EPQ)}$' represent the best optimal solution and the position vector of the emperor penguin respectively. '$S_{F(EPQ)}$' is the social factor of an emperor penguin attributing towards the movement in the direction of the best optimal search agent. At this juncture, the collision prevention factors '$\overrightarrow{C_{PV(1)}}$' and '$\overrightarrow{C_{PV(2)}}$' are determined based on Eqs. (9) and (10).

$$\overrightarrow{C_{PV(1)}} = \left(M_P \times \left(T^C_{\text{Huddle}} + ACC_{P-\text{Grid}}\right)\right) \times \left(\text{Rand}\left(\right) - T^C_{\text{Huddle}}\right) \tag{9}$$

$$\overrightarrow{C_{PV(2)}} = \text{Rand}\left(\right) \tag{10}$$

With '$ACC_{P-\text{Grid}}$' determined based on Eq. (11),

$$ACC_{P-\text{Grid}} = Abs\left(S_{\text{Best}(EPQ)} - S_{PVC(EPQ)}\right) \tag{11}$$

In this scenario, '$ACC_{P-\text{Grid}}$' is the accuracy of the polygon grid determined based on the deviation between the random function (Rand() ∈ [0, 1]) and emperor penguins. Moreover, the parameter of movement (M_P) and the huddle temperature (T^C_{Huddle}) are sustained for estimating the deviation between the search agents for preventing collisions. In the proposed scheme, the value of 'M_P' is set to 2. In addition, the value of '$S_{F(EPQ)}\left(\overrightarrow{C_{PV(1)}}\right)$' is computed based on Eq. (12).

$$S_{F(EPQ)}\left(\overrightarrow{C_{PV(1)}}\right) = \left(\sqrt{C_{P\left(\text{Exploitation}\right.}\left(e^{\frac{\text{Iter}_{\text{Current}}}{C_{P(\text{Exploration})}}} - e^{\text{Iter}_{\text{Current}}}\right)}\right)^2 \tag{12}$$

where '$C_{P(\text{Exploitation})}$' and '$C_{P(\text{Exploration})}$' are the parameters of control used for achieving better exploitation and exploration degree with 'e' as the exponential function. The value of '$C_{P(\text{Exploitation})}$' and '$C_{P(\text{Exploration})}$' ranges between [2, 3] and [1.5, 2] respectively.

4.4 *Effective Mover Relocation*

In this final phase, the emperor penguin positons are individually updated based on the determined superior optimal solution representing a mover. This mover aids in changing positions of the remaining search agents in the search space through which vacation with respect to current position is attained based on Eq. (13).

$$S_{PVC(EPQ)}\left(\text{Iter}_{\text{Current}} + 1\right) = S_{PVC(EPQ)}\left(\left(\text{Iter}_{\text{Current}}\right) - \overrightarrow{C_{PV(1)}} \times \overrightarrow{D_{IEPQ}}\right) \tag{13}$$

In this process of iteration, the emperor penguins' huddling characteristics is recalculated based on the relocated mover position. In the proposed scheme, the evaluation of multi-constrained fitness evaluation is achieved using the objectives of reduced delay, maximum node energy, minimum node distance and *LLT*. This algorithm executes until the termination condition of determining the reliable nodes along the multicast path or the number of iterations utilized for implementation is achieved.

5 Simulation Results and Discussion

The simulation results of the proposed EPOA-MT-MAODV and the benchmarked Cuckoo Search-Multicast Tree-Multicast Ad hoc On-Demand Vector (CS-MT-MAODV), Power-aware Dual-Tree-based Multicast Routing Protocol (PDTMRP) and Reliable and Energy-Aware Multicast Ad hoc On-demand Distance Vector (REA-MAODV) schemes are carried out using MATLAB installed in the system using Windows 10 OS and CPU 2.16 GHz processor evaluated based on energy consumption and LLT for different number of mobile nodes and mobility speeds.

Initially, Figs. 3 and 4 present the predominance of the proposed EPOA-MT-MAODV and the benchmarked CS-MT-MAODV, PDTMRP and REA-MAODV schemes evaluated based on energy consumption and LLT for varying number of mobile nodes respectively. The energy consumption of the proposed EPOA-MT-MAODV scheme is visualized to be improved on par with the baselines even with increase in the number of mobile nodes in the network. This predominance in energy consumption is mainly due to the adaptive huddle factor used in EOPA strategy that aids in routing the data packets only through reliable nodes in the routing path. This reliable identification of reliable nodes plays an anchor role in maintaining

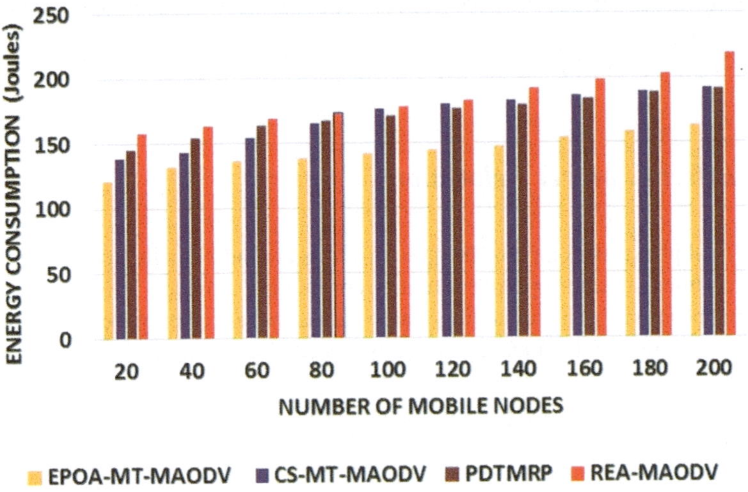

Fig. 3 Energy consumption of the proposed EPOA-MT-MAODV scheme for varying number of mobile nodes

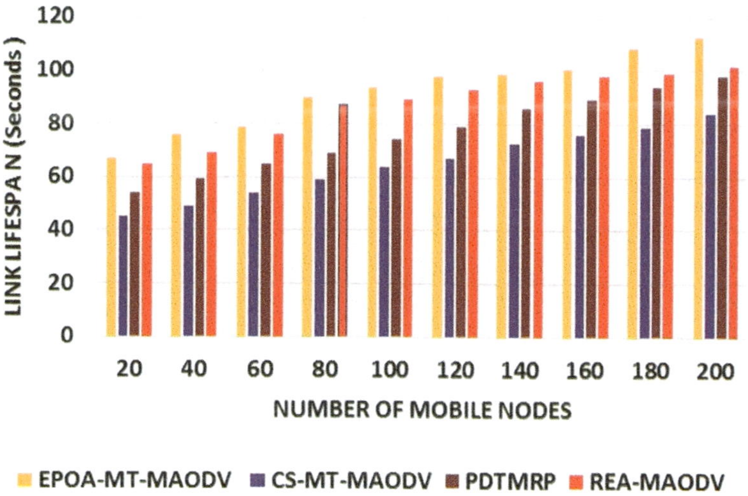

Fig. 4 Link lifespan time of the proposed EPOA-MT-MAODV scheme for varying number of mobile nodes

the LLT of the network up to a considerable level. The energy consumption of the proposed EPOA-MT-MAODV for varying number of mobile nodes is minimized by 7.21%, 8.94% and 10.68% when compared to the benchmarked CS-MT-MAODV, PDTMRP and REA-MAODV schemes. The LLT of the proposed EPOA-MT- for varying number of mobile nodes is improved by 6.54%, 7.68% and 8.42% when compared to the benchmarked CS-MT-MAODV, PDTMRP and REA-MAODV schemes.

In addition, Figs. 5 and 6 demonstrate the potential of the proposed EPOA-MT-MAODV and the benchmarked CS-MT-MAODV, PDTMRP and REA-MAODV schemes evaluated based on energy consumption and LLT for varying mobility speeds of mobile nodes. The energy consumption of the proposed EPOA-MT-MAODV scheme for varying mobility speeds is identified to get enhanced when compared to the baseline schemes, even with increase in the rate of mobility of nodes in the network. The energy consumption of the proposed EPOA-MT-MAODV is considerably reduced with the use of relocation factor that constantly increases the possibility of determining the optimal nodes in the network. The energy consumption of the proposed EPOA-MT-MAODV for varying mobility speeds is reduced by 6.94%, 7.86% and 8.94% when compared to the benchmarked CS-MT-MAODV, PDTMRP and REA-MAODV schemes. The LLT of the proposed EPOA-MT-MAODV for varying mobility speeds is enhanced by 7.42%, 8.64% and 9.28% when compared to the benchmarked CS-MT-MAODV, PDTMRP and REA-MAODV schemes.

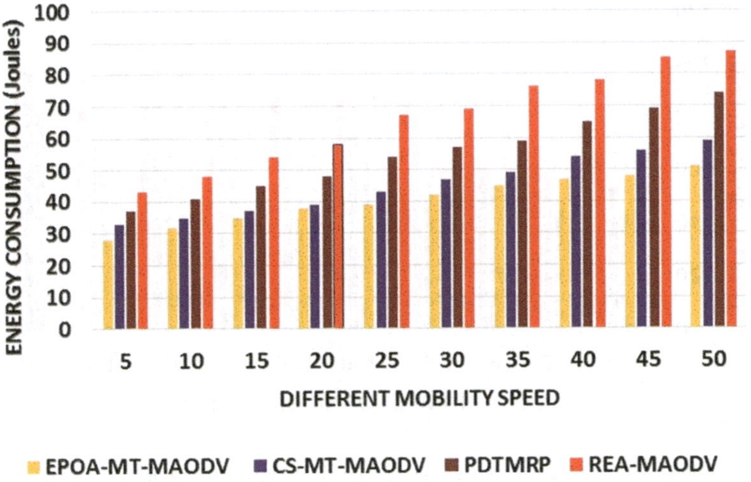

Fig. 5 Energy consumption of the proposed EPOA-MT-MAODV scheme for varying mobility speeds

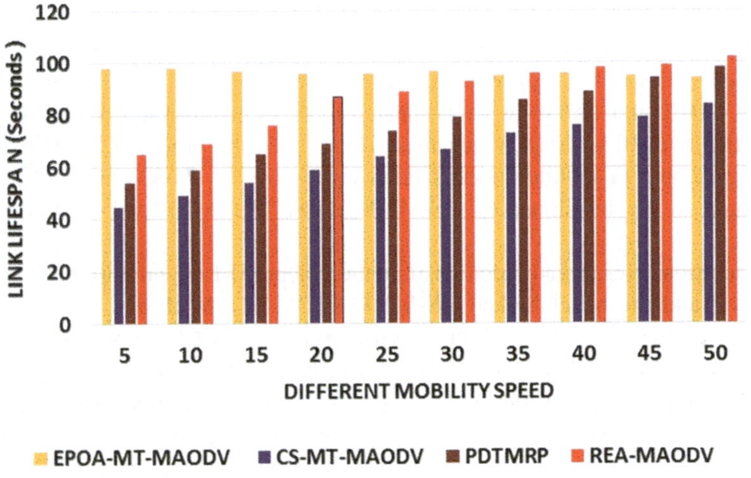

Fig. 6 Link lifespan time of the proposed EPOA-MT-MAODV scheme for varying mobility speeds

6 Conclusion

In this chapter, Emperor Penguin Optimization Algorithm and M-Tree-based Multicast Ad hoc On-demand Distance Vector Routing (EPOA-MT-MAODV) protocol is proposed for efficient selection of multicast routes and enhancing network lifetime. This EPOA-MT-MAODV protocol is proposed with benefits of huddle factor and mover relocation for sustaining balance between the rate of exploitation and exploration during optimal multicast tree construction process. It computes the fitness function formulated based on the factors of delay, minimum distance, link stability and energy for optimal selection of optimal routes facilitating reliable multicast routing process. The simulation experiments conducted through MATLAB prove that the energy consumption of the proposed EPOA-MT-MAODV for varying mobility speeds are significantly reduced by 6.94%, 7.86% and 8.94% when compared to the benchmarked CS-MT-MAODV, PDTMRP and REA-MAODV schemes. The results also prove that the LLT of the proposed EPOA-MT- MAODV for varying number of mobile nodes is improved by 6.54%, 7.68% and 8.42% when compared to the benchmarked CS-MT-MAODV, PDTMRP and REA-MAODV schemes.

References

1. Priya, M. D., & Priyanka, P. (2015). PPCLSS: Probabilistic prediction coefficient link stability scheme based routing in MANETs. *International Journal of Computer Science & Engineering Technology, 6*(4), 246–256.
2. Priya, M. D., Aishwarya, R., Anushya, S., & Keerthana, S. (2017). Detection and avoidance of malicious nodes in MANETs. *International Journal of Scientific Research in Computer Science, Engineering and Information Technology, 2*(2), 434–441.
3. Maharaja, S., Jeyalakshmi, R., Kanna, S. A. V., & Priya, M. D. (2019). Secured routing in Mobile ad hoc networks (MANETs). *International Journal of Scientific Research in Computer Science, Engineering and Information Technology, 5*(2), 277–289.
4. Priya, M. D., Janakiraman, S., Sandhya, G., & Aishwaryalakshmi, G. (2019). Efficient pre-authentication scheme for inter-ASN handover in high mobility MANET. *Wireless Networks, 27*, 1–15.
5. Yen, Y. S., Chan, Y. K., Chao, H. C., & Park, J. H. (2008). A genetic algorithm for energy-efficient based multicast routing on MANETs. *Computer Communications, 31*(10), 2632–2641.
6. Astier, E., Hafid, A., & Benslimane, A. (2009). Energy and mobility aware clustering technique for multicast routing protocols in wireless ad hoc networks. In *2009 IEEE wireless communications and networking conference* (pp. 1–6).
7. Hu, C. C. (2010). Bandwidth-satisfied multicast trees in large-scale ad-hoc networks. *Wireless Networks, 16*(3), 829–849.
8. Jabbehdari, S., Shamaei, M., & Darehshoorzadeh, A. (2010). IQoS-ODMRP: A novel routing protocol considering QoS parameter in MANET. In *IEEE symposium on industrial electronics and applications* (pp. 126–130).
9. Hwang, S. F., Su, Y. Y., Lu, K. H., & Dow, C. R. (2011). A cluster-based approach for efficient multi-source multicasting in MANETs. *Wireless Personal Communications, 57*(2), 255–275.
10. Santhi, G., & Nachiappan, A. (2011). Fuzzy cost based multicast routing for mobile ad-hoc networks with improved QoS. In *International conference on digital image processing and information technology* (pp. 429–437). Berlin, Heidelberg: Springer.
11. Biradar, R. C., & Manvi, S. S. (2011). Agent-driven backbone ring-based reliable multicast routing in mobile ad hoc networks. *IET Communications, 5*(2), 172–189.
12. Kharraz, M. A., Sarbazi-Azad, H., & Zomaya, A. Y. (2012). On-demand multicast routing protocol with efficient route discovery. *Journal of Network and Computer Applications, 35*(3), 942–950.
13. Wang, N. C. (2012). Power-aware dual-tree-based multicast routing protocol for mobile ad hoc networks. *IET Communications, 6*(7), 724–732.

14. Kim, M., Choo, H., Mutka, M. W., Lim, H. J., & Park, K. (2013). On QoS multicast routing algorithms using k-minimum Steiner trees. *Information Sciences, 238*, 190–204.
15. Li, X., Liu, T., Liu, Y., & Tang, Y. (2014). Optimized multicast routing algorithm based on tree structure in MANETs. *China Communications, 11*(2), 90–99.
16. Kumar, Y. A., & Sachin, T. (2016). A tree based multicast routing protocol using reliable neighbor node for wireless mobile ad-hoc networks. In *Proceedings of the 4th international conference on Frontiers in intelligent computing: Theory and applications* (pp. 455–465). New Delhi: Springer.
17. Yadav, A. K., & Tripathi, S. (2017). QMRPRNS: Design of QoS multicast routing protocol using reliable node selection scheme for MANETs. *Peer-to-Peer Networking and Applications, 10*(4), 897–909.
18. Tavizi, A., & Ghaffari, A. (2018). Tree-based reliable and energy-aware multicast routing protocol for mobile ad hoc networks. *The Journal of Supercomputing, 74*(11), 6310–6332.
19. Hassan, M. H., Mostafa, S. A., Mohammed, M. A., Ibrahim, D. A., Khalaf, B. A., & Al-Khaleefa, A. S. (2019). Integrating African Buffalo optimization algorithm in AODV routing protocol for improving the QoS of MANET. *Journal of Southwest Jiaotong University, 54*(3).
20. Papanna, N., Reddy, A. R. M., & Seetha, M. (2019). EELAM: Energy efficient lifetime aware multicast route selection for mobile ad hoc networks. *Applied Computing and Informatics, 15*(2), 120–128.

Fog-Assisted Real-Time Coronary Heart Disease Risk Detection in IoT-Based Healthcare System

L. Jubair Ahmed, B. Anishfathima, B. Gokulavasan, and M. Mahaboob

Abstract Advances in healthcare systems are helpful in the diagnosis and treatment of extremely critical diseases. Continuous monitoring of individuals leads to huge amount of medical data, and new technology solutions are highly essential to handle the generated data. Fog computing approach is proposed in this chapter to develop an efficient wearable device for wireless healthcare monitoring. Fog computing proves to be better than other remote health monitoring methods because of its fast decision-making capability and delivery of simple notifications to users. Centralized cloud server is utilized in fog computing framework for performing complex computations. In this work, machine learning-based Coronary Heart Disease (CHD) risk detection is carried out with fog manipulation technique. Machine learning-based CHD risk assessment is performed on HRV feature-extracted ECG signal, blood pressure, glucose and cholesterol level. Wavelets are applied in this work for ECG signal feature extraction and identifying heart rate-related parameters. Various machine learning algorithms such as decision tree classifier, SVM, ANFIS and KNN are utilized for classification. ANFIS classification is found to be effective in CHD risk prediction; it categorizes the given individual data into normal and CHD risky. Based on this work, it is possible to provide warning to the individuals to correct their food habits and lifestyle.

Keywords Fog computing · Internet of things (IoT) · Machine learning · Coronary heart disease (CHD) · Wireless healthcare system · Heart rate variability (HRV)

L. J. Ahmed (✉) · B. Gokulavasan · M. Mahaboob
Sri Eshwar College of Engineering, Coimbatore, Tamil Nadu, India

B. Anishfathima
Sri Krishna College of Engineering and Technology, Coimbatore, Tamil Nadu, India
e-mail: anishfathimab@skcet.ac.in

A. Haldorai et al. (eds.), *3rd EAI International Conference on Big Data Innovation for Sustainable Cognitive Computing*, EAI/Springer Innovations in Communication and Computing, https://doi.org/10.1007/978-3-030-78750-9_8

1 Introduction

Nowadays, healthcare technologies and biomedical equipment have grown rapidly due to advances in semiconductor industry. Though there are technological improvements in the modern medical equipment available in healthcare centres, it is not a simple task to develop affordable, low complexity and high-quality wearable devices for ECG tele-monitoring [1]. Mobile healthcare monitoring systems are highly popular in recent times due to the size of wearable device and their real-time monitoring capability. In continuous monitoring, data of individuals need to be analysed effectively to recognize critical health conditions [2]. Mobile-based monitoring encounters huge challenges in developing low complexity power efficient systems. Cloud computing and fog computing emerge as major computing paradigms in the era of Internet of things (IoT), where it is possible to utilize bulky information saving options along with huge manipulation facilities efficiently [3]. Figure 1 highlights cloud computing-related telemedicine where hospital service system, physicians and users are connected through IoT. Mobile cloud computing (MCC) allows the healthcare centres to store numerous medical records that are generated in daily clinical practices. The cloud server is accessed by healthcare centres while performing data analysis and decision making [4].

In MCC, cloud computing services can be utilized in intensive computation-based applications in addition to huge data storage. Cloud computing power has been utilized in electroencephalograph (EEG) signal feature extraction and analysis in the detection of brain-related diseases [5]. MCC techniques find applications in computation-intensive ECG tele-monitoring applications. Cloud computing facilitates the minimum local processing of signals in IoT-based healthcare devices. However, there are few issues in quality of service while using cloud framework in real-time continuous monitoring. Fog computing has gained popularity among researchers in healthcare applications to enhance security and save bandwidth. It provides energy efficiency in sensor nodes and enables distributed data storage and data compression. A smart personalized healthcare monitor has been developed

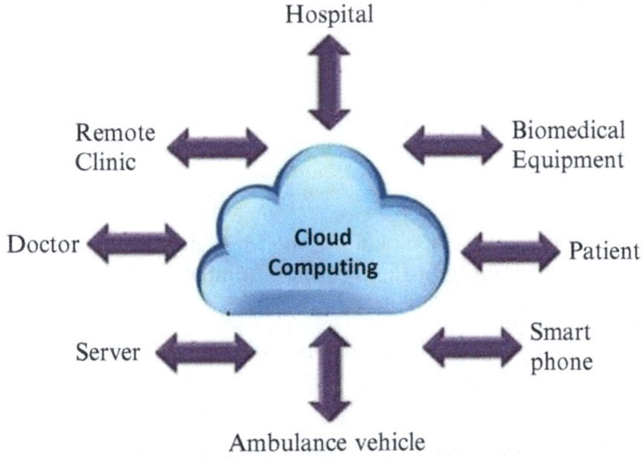

Fig. 1 Cloud computing model for digital healthcare

for ECG signal monitoring from remote locations that utilizes cloud services for medical data storage and analysis [6]. In this chapter, ECG feature extraction analysis and machine learning-based classification are carried out for coronary heart disease (CHD) risk detection that utilizes fog computing architecture.

CHD occurs whenever there is a blood circulation flow disturbance to coronary muscles. The blood supply reduction is predominantly due to cholesterol deposits in the blood vessels. The reduction of blood circulation to the pumping organ distracts the respiratory gases volume along with nourishment supply that affects regular working mechanism. Reduction of O_2 with essential nourishment components end up in heartache leading to heart attack to the affected circle [6]. Since CHD causes most of the heart-related problems and heart attack, CHD risk assessment methods are vital to avoid serious heart diseases [7]. Usual health indicators for evaluating CHD risk detection include age, smoking habits and medical history. Early identification of cardiovascular diseases is essential to proceed with the proper treatment [7]. Blood pressure and heart rate variability (HRV) parameters were evaluated to detect CHD possibilities and mortality using Dublin survey [8]. The study suggested the need for continuous real-time monitoring of ECG signal for preliminary level CHD detection. Besides blood pressure and HRV-based prediction, tobacco usage has also been taken for the evaluation of ventricular arrhythmias.

Electrocardiogram (ECG) morphological investigation and HRV-based analysis have led many researchers into in-depth analysis of ECG signal features. The R-R interval of ECG signal is measured to examine heart beat duration in millisecond units (ms). Followed by heart rate in bpm is measured with a formulation = (60000)/(RR interval). If any modifications seen in R-R interval shows abnormal heart rate, then the R-R intervals will be studied in spectrum with transforms like FFT, STFT as well as wavelet transforms [9]. HRV features are extracted in both time domain and frequency domain for analysing cardiovascular activities [10–12]. In human HRV, there are three main categories, namely, very low frequency (VLF, i.e. less than 0.04 Hz), low frequency (LF, i.e. between 0.04 and 0.15 Hz) and high frequency (HF, i.e. between 0.15 and 0.5 Hz). In machine learning-based classifiers, observed time and frequency domain metrics were utilized for training classifier.

Wavelets have been largely deployed for information extraction process in many biomedical signals and images. Since all the wavelets are formed by shifting and scaling of mother wavelet, mother wavelet selection is crucial in many applications [13–14]. Machine learning algorithms are being utilized in many healthcare applications for detection of diseases by performing classification. Artificial neural network is used in stroke diagnosis healthcare system by applying age, gender and medical history [15].Though there are many works in the detection of acute stroke and heart diseases, it is necessary to predict acute stroke to avoid sudden death. CHD risk detection is highly essential for acute stroke prediction. Artificial neural networks are proposed for ECG signal analysis where the complex relationship has been established between inputs and outputs [13]. In addition to neural networks, support vector machines, fuzzy logic and genetic algorithms are used for ECG signal analysis and classification. In few scenarios, neural networks are combined with fuzzy logic for obtaining desired results. In adaptive neuro fuzzy inference system, a back propagation learning algorithm of neural network is combined with fuzzy inference system [14].

In ECG signal processing, identifying the exact position of the QRS complex helps to detect heart rate variability (HRV) parameters [16]. Many cardiovascular-related diseases are identified by analysing the ECG signal parameters such as QRS complex, R-R interval, P wave and QT interval dispersion [17–18]. Though ECG signal is a reliable indicator to identify acute stroke and heart-related diseases, ECG signal parameters are not studied in depth in CHD threat detection. Results were acquired and related with comparable techniques. The importance of the proposed work is listed as follows: (i) personalized warning can be given to the individuals to correct their food habits and lifestyle; (ii) elderly people can be informed about heart functionality to take necessary precautions; (iii) cause of heart diseases can be identified by clinicians for better treatment. There are four sections in the chapter including the introduction part. The second section presents the ECG signal feature extraction using wavelets, ANFIS evaluation of HRV signal and CHD risk detection. The third section represents the acquired end product and advantages when related with already existing mechanisms. To conclude, summary of our observations are presented.

2 Architecture and Methodology of the System

Our proposed system comprises of fog manipulation technique, feature classification of HRV and CHD threat detection that are explained as follows. The goals of the work are: (i) to collect large volume of data using wearable devices while monitoring ECG, blood pressure, glucose and cholesterol level; (ii) to develop principal component analysis-based ECG signal feature extraction; (iii) to develop the fog computing framework for storing and analysing the collected data from monitoring devices; iv) To apply machine learning algorithm for CHD risk detection based on extracted ECG signal features and medical data.

2.1 Fog Computing Model

The individuals' data can be collected from hospitals that belong to different age group, gender, background and patients with heart-related diseases. MIT-BIH database ECG signal analysis can be used for testing. ECG signal can also be acquired using NI DAQ and wearable devices for better testing in real-time scenario. These collected data are sent to fog environment without any processing to reduce the computational burden of wearable devices. Complex quality analysis and stroke prediction are performed using fog computing framework. Fog computing framework comprises of sensor layer, fog computing layer with cloud computation

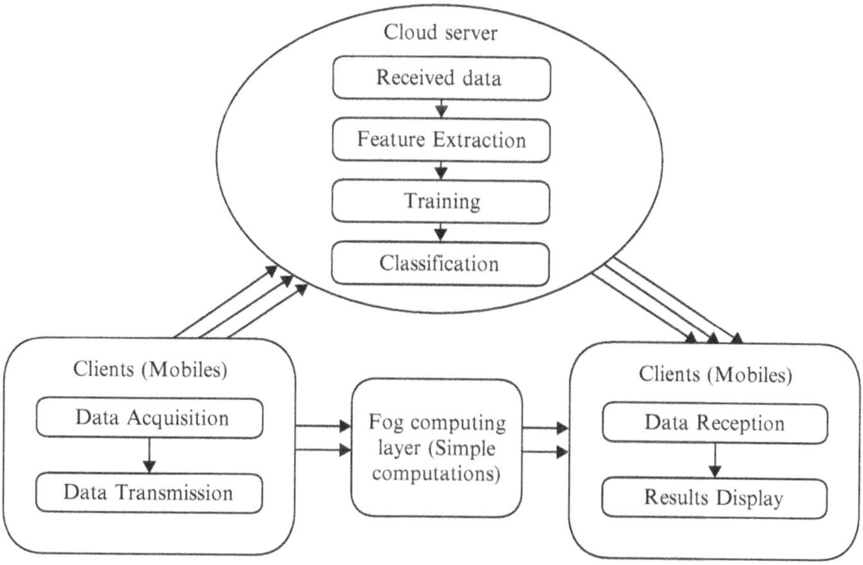

Fig. 2 Fog computing model

level. Figure 2 refers to the method with the prescribed cloud computation method. Wearable device dataset are collected by sensor layer and they are sent to fog computing layer for signal quality assessment and to communicate few simple decisions. Heavy computations are performed in cloud computing layer, and the results are sent back to the end user.

2.2 *HRV Feature Extraction*

The collected data have to be sent to the fog environment for machine learning-based CHD risk detection. Blood pressure, glucose level, cholesterol level can be sent without any preprocessing. However, preprocessing and feature extraction plays a major role in ECG signal analysis. A complete data analysis on ECG signal with signal processing techniques for heart rate feature extraction is performed. Time and spectrum investigations were examined with acquired HRV of ECG. Many wavelets were utilized for HRV feature extraction where wavelets and their results were finalized with number of coefficients and decomposition level. Coif wavelet was used in this work producing good HRV metric acquisition from the ECG, because linear phase characteristics are much better than Daubechies wavelet and Spline wavelet [9]. For R-peak estimation, the number of wavelets was exploited. Coif wavelet provided good outcomes to detect QRS peak in ECG signals [11]. Practically, R-R mid time evaluation model having timing resolution ± 1ms is used with correlation technique.

2.3 CHD Threat Detection

The collected and feature extracted data are stored as training data and testing data. Usually 80% of the data can be trained to obtain good classification results. The classification of individuals data can be done as normal and CHD risky. Machine learning-based CHD risk classification is carried out in this stage by the following steps: (i) feature extraction from the quality-assessed ECG signal; (ii) applying various physiological data and QRS feature-extracted ECG signals; (iii) machine learning-based classification for acute stroke prediction into normal, low-stroke risk and high-stroke risk. Figure 3 describes the machine learning-based CHD risk classification.

CHD risk detection in fog computing framework is carried out using the above machine learning-based process flow. The prototype can be developed as a wearable device for better CHD risk detection. The prototype can be checked with the combination of wearable devices and fog computing-assisted machine learning

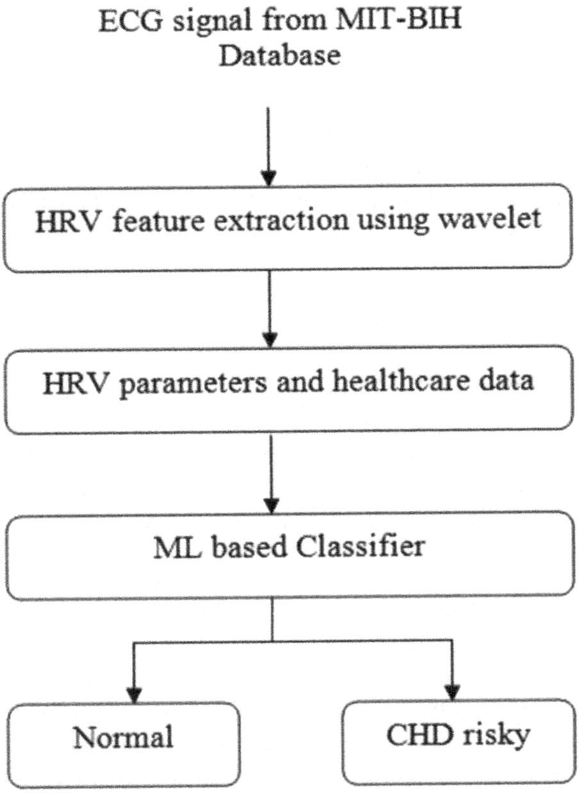

Fig. 3 Machine learning-based CHD risk classification

framework. The classification accuracy of CHD risk detection is determined using specificity, sensitivity and total classification accuracy.

3 Results and Discussion

Results of HRV feature extraction and machine learning-based classification are presented and their significance is discussed in this section. Patient data can be collected from hospitals that belong to different age group, gender, background and patients with heart-related diseases. MIT-BIH database are also applied for better training of the classifier. Matlab 2015a and NI DAQ are used in this work for analysis and better testing in real-time scenario. These collected data are sent to fog environment without any processing to reduce the computational burden of wearable devices. Data analysis is performed with the sample size of 160 in the cloud server under the fog computing environment. Wavelet-based HRV feature extraction is performed for generating many frequency and time domain metrics. National Instruments Biomedical equipment was used for evaluating these metrics. Hence, collected data are stored as training data and testing data. Usually, 80% of the data can be trained to obtain good classification results. The classification of individuals data can be done as normal, CHD risky. Machine learning-based quality analysis and risk classification are performed.

Observed ECG signal data from MIT-BIH database and real-time recorded signals are given to wavelet-based HR feature extraction process. R-peak detection is the major step towards HRV feature extraction. Coif wavelet and biorthogonal wavelets are used to detect the possible R-peaks in one healthy individual ECG data. The obtained result is shown in Fig. 4. From the obtained results, biorthogonal wavelet provided the beat rate of 129 whereas Coif wavelet provided the beat rate of 74. Since the usual human heart rate ranges from 60 to 100 bpm, it is concluded that it cab be analyzed with Coif wavelet.

National instruments (NI) health equipment was used for examining spectrum domain values. The obtained values were averaged for studying the threats utilizing arrhythmic beat classification technique. From Table 1, it is clear that all cases of HR mean ranges between 70 bpm and 90 bpm. Various numerical metrics like HR average (mean), HR (std. deviation), RR average, RR SD, R-MSSD, NN-50 and TINN (Triangular Index NN) were analysed (Fig. 5).

The observed RR metrics were eventually studied with the help of Fuzzy Inference System (FIS) in order to detect CHD threat. Table 2 represents the rule base based on the 25 fuzzy rules.

The proposed system worked with 128 training datasets alongside 1000 epochs (80%) that was utilized to study the ANFIS classifier. Preliminary phase of training is crucial for obtaining better classification accuracy. After training the data samples, 32 test cases out of 40 samples (80%) were applied for verifying the classification exactness of this machine learning prototype. Confusion matrix in Table 3 is used to determine the classification accuracy of the proposed work. The ANFIS

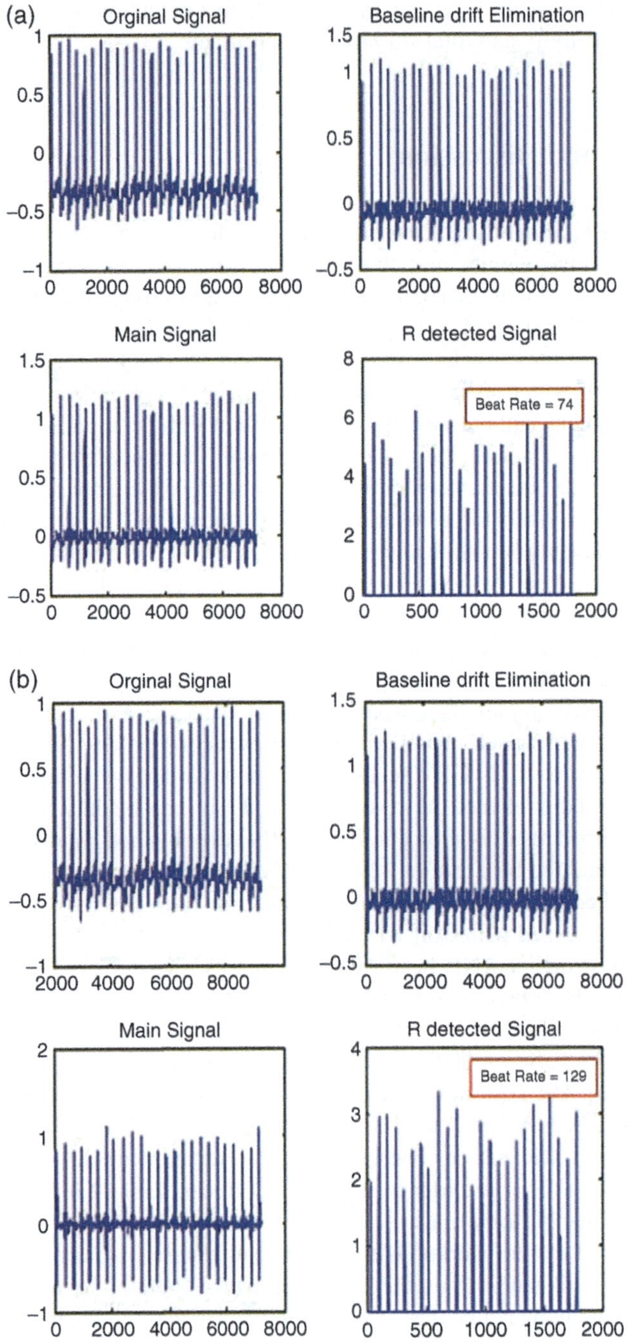

Fig. 4 Analysis of R-peak using (**a**) Biorthogonal wavelet, (**b**) Coif wavelet

Table 1 Obtained statistical parameters of ECG data

Subject	Avg. RR	Std. RR	Avg. HR	Std. HR	R-MSSD Metric	NN-50 Metric	pNN-50 Metric	TINN Metric
1	774	180	75	9.1	270	9	22	80
2	748	30	75	3.6	39	10	24	120
3	748	130	76	3.4	48	1	2.4	89
4	773	133	75	2.7	28	20	41	57
5	771	121	74	2.4	28	5	19	115
6	771	19	76	10.2	250	6	19	77
7	679	160	74	5.2	47	1	22	38
8	620	144	73	3.6	39	17	31	127
9	590	110	75	9.8	230	6	20	116
10	632	95	70	8.7	160	12	33	49
11	679	158	78	10	250	9	10.4	97
12	701	25	73	3.6	25	14	28	124
13	718	126	85	20.6	42	9	22	80
14	704	128	80	6.2	28	10	24	120
15	749	143	72	51.7	73	1	2.4	89

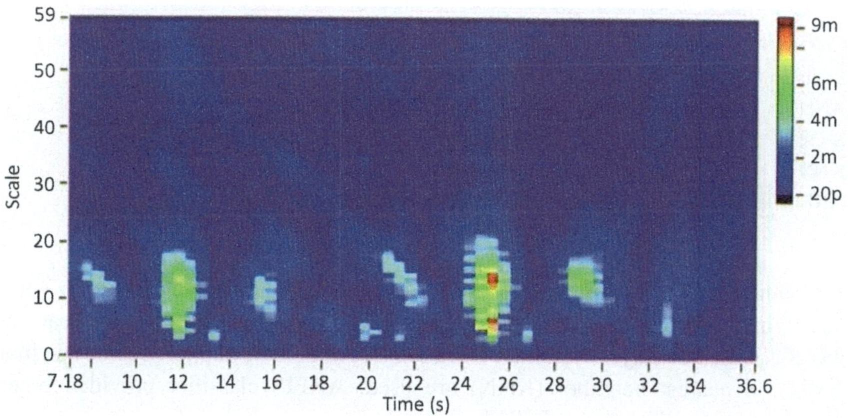

Fig. 5 Obtained wavelet coefficients using NI biomedical kit

classifier performance is analysed with the help of the following metrics: specificity, sensitivity and total classification accuracy. The statistical parameter values such as sensitivity, specificity and total classification accuracy are derived from Table 3. Table 4 shows the attained specificity, sensitivity and total classification accuracy of ANFIS classifier.

Formula for finding total classification accuracy (TCA):

$$TCA = \frac{\text{(sum of the value of sensitivity with the value of specificity)}}{2}$$

Table 2 CHD risk-based rule base

RR mean	RR standard deviation				
	ES	S	A	L	EL
ES	A	A	L	EL	EL
S	A	L	L	EL	EL
A	L	A	A	A	L
L	S	S	A	A	L
EL	ES	ES	L	A	S

ES extremely small, *S* small, *A* average, *L* large, *EL* extremely large

Table 3 Confusion matrix

Desired result	Output result	
	HRV: normal	HRV: CHD risk
HRV: normal	158	2
HRV: CHD risk	1	158

Table 4 Statistical parameter values

Specificity (%)	Sensitivity (%)	Total classification accuracy (%)
100	97.5	98.75

Table 5 Evaluation and results of different classification

Classification method	Bioindicators and features	Accuracy of classification (%)
Decision tree classifier	ECG, blood volume, pulse	88.25%
ANN	ECG, blood volume, cholesterol level	89.75%
SVM	Blood volume, ECG, temperature	91.49%
KNN	HRV, blood volume	96.25%
ANFISclassifier	HRV, age, diabetes, cholesterol level	99.38%

Various machine learning models are related as shown with Table 5 while classifying subjects into CHD risky vs. normal. Though many machine learning classifiers such as Bayes classifier, Decision tree classifier, Support vector machine (SVM), K-nearest neighbor (KNN) are used, ANFIS classifier provided better classification for the applied HRV features. In many machine learning approaches, different biomedical data such as peaks of ECG signal, capacity of blood, rate of pulse, temperature as well as HRV metrics were considered. From the obtained results, it is observed that the maximum classification accuracy is 96.88%. ANFIS classifier yields the extreme 99.38% accuracy after sorting as normal vs. CHD risky.

4 Conclusion

This chapter suggested a fog computing-assisted IoT framework for telemedicine applications. Machine learning-based CHD risk detection is proposed using extracted HRV features. Fog computing layer is utilized for fast and reliable decision making. Machine learning-based heavy computations are performed in the cloud server. In various machine learning approaches, biomedical data such as peaks of ECG signal, capacity of blood, rate of pulse, temperature as well as HRV metrics are considered for CHD risk detection. From the obtained results, it is observed that the maximum classification accuracy is 96.88%. However, ANFIS classifier yields the extreme 99.38% accuracy after sorting as normal vs. CHD risky.

References

1. Varshney, U. (2009). Pervasive healthcare computing: MR/HER. In *Wireless and health monitoring*. New York, NY: Springer Verlag.
2. Wang, X., Gui, Q., Liu, B., Jin, Z., & Chen, Y. (2014 March). Enabling smart personalized healthcare: A hybrid mobile-cloud approach for ECG telemonitoring. *IEEE Journal of Biomedical and Health Informatics, 18*(3), 739–745.
3. Wang, X., Gui, Q., Liu, B., Chen, Y., & Jin, Z. (2013 April). Leveraging mobile cloud for telemedicine: A performance study in medical monitoring. In *Proceedings of the39thnortheast bioengineering conference* (pp. 49–50). IEEE.
4. Hsieh, J. C., & Hsu, M. W. (2012 July). A cloud computing based 12-lead ECG telemedicine service. *BMC Medical Informatics and Decision Making, 12*(1), 77.
5. Shen, C. P., Chen, W. H., Chen, J. M., Hsu, K. P., Lin, J. W., Chiu, M. J., Chen, C. H., & Lai, F. (2010 November). Bio-signal analysis system design with support vector machine based on cloud computing service architecture. In *Proceedings of the international conference on IEEE engineering in medicine and biology. Society* (pp. 1421–1424).
6. Wang, X., Gui, Q., Liu, B., Jin, Z., & Chen, Y. (2014). Enabling smart personalized healthcare:A hybrid mobile-cloud approach for ECG telemonitoring. *IEEE Journal of Biomedical and Health Informatics, 18*(3), 739–745.
7. McMahan, C., Gidding, S., Fayad, Z., Zieske, A., Malcom, G., & Tracy, R. (2005 April). Risk scores predict atherosclerotic lesions in young people. *Archives of Internal Medicine, 165*(8), 883–890.
8. Übeyli, E. D. (2008 March). Adaptive neuro-fuzzy inference system employing wavelet coefficients for detection of ophthalmic arterial disorders. *Expert Systems with Applications, 34*(3), 2201–2209.
9. Kumar, M., Weippert, M., Vilbrandt, R., Kreuzfeld, S., & Stoll, R. (2007 May). Fuzzy evaluation of heart rate signals for mental stress assessment. *IEEE Transactions on Fuzzy Systems, 15*(5), 791–808.
10. Wiggins, M., Saad, A., Litt, B., & Vachtsevanos, G. (2008 January). Evolving a Bayesian classifier for ECG based age classification in medical applications. *Applied Soft Computing, 8*(1), 599–608.
11. Satheeskumaran, S., & Sabrigiriraj, M. (2014 August). A new LMS based noise removal and DWT based R-peak detection in ECG signal for biotelemetry applications. *National Academy Science Letters, 37*(4), 341–349.
12. Wojtasik, Z., Jaworski, W., Kuźmicz, A., Wielgus, A. W., & Sarna, D. (2004 March). Fuzzy logic controller for rate-adaptive heart pacemaker. *Applied Soft Computing, 4*(3), 259–270.
13. Erkaymaz, H., Ozer, M., & Orak, İ. M. (2015). Detection of directional eye movements based on the electrooculogram signals through an artificial neural network. *Chaos, Solitons & Fractals, 77*, 225–229.
14. Venkatesan, C., Karthigaikumar, P., & Satheeskumaran, S. (2018). Mobile cloud computing for ECG telemonitoring and real-time coronary heart disease risk detection. *Biomedical Signal Processing and Control, 44*, 138–145.

15. Karaca, Y., Moonis, M., Zhang, Y.-D., & Gezgez, C. (2019). Mobile cloud computing based stroke healthcare system. *International Journal of Information Management, 45*, 250–261.
16. Deepu, C. J., Zhang, X., Heng, C. H., & Lian, Y. (2016 December). A 3- lead ECG-onchip with QRS detection and lossless compression for wireless sensors. *IEEE Transactions on Circuits and Systems II: Express Briefs, 63*(12), 1151–1155.
17. Chen, Y.-P., et al. (2015 January). An injectable 64 nW ECG mixed- signal SoC in 65 nm for arrhythmia monitoring. *IEEE Solid-State Circuits, 50*(1), 375–390.
18. Lederman, Y. S., Balucani, C., Steinberg, L. R., Philip, C., Lazar, J. M., Weedon, J., Mirchandani, G., et al. (2019). Does the magnitude of the electrocardiogram QT interval dispersion predict stroke outcome? *Journal of Stroke and Cerebrovascular Diseases, 28*(1), 44–48.

Food Demand Forecast for Online Food Delivery Service Using CatBoost Model

Ansh Pujara, V. Pattabiraman, and R. Parvathi

Abstract Online food ordering has been proven a great source for businesses from a wide range of sectors. By using an online food ordering system, you can get your food to be delivered to your door without consuming much time. For businesses operating in the food industry including restaurants, agriculture, and many others, accurate forecast is of crucial importance because of the unpredictable demand pattern. In several studies, the choice of an appropriate forecasting model remains a concerning point. In this context, this research aims to analyse the performance of the CatBoost Gradient boosting model for the prediction of the amount of raw materials required for a meal delivery company that operates in multiple cities having multiple centres.

Keywords Food demand · CatBoost · Demand forecast

1 Introduction

Demand forecasting is an integral component of any rising online business. Without proper demand forecasting techniques in place, it can be nearly impossible to have the required amount of stock available at any given time [1]. A major facet of businesses today is the notion of supply chain integration, whereby resources are combined to provide value to the end consumer and where all the upstream firms realize the importance of integration.

An online food delivery service has to face several difficulties like dealing with a lot of perishable raw materials, which makes it perhaps more essential for such a company to accurately forecast daily and weekly demand [2]. The task is to predict the demand of raw materials for 10 weeks in future using the historical data for a product-centre combination of the past 145 weeks.

A. Pujara · V. Pattabiraman · R. Parvathi (✉)
Vellore Institute of Technology, Chennai, Tamil Nadu, India
e-mail: ansh.pujara2018@vitstudent.ac.in; pattabiraman.v@vit.ac.in; parvathi.r@vit.ac.in

A. Haldorai et al. (eds.), *3rd EAI International Conference on Big Data Innovation for Sustainable Cognitive Computing*, EAI/Springer Innovations in Communication and Computing, https://doi.org/10.1007/978-3-030-78750-9_9

Prediction of the type of raw material required at a particular delivery centre/branch depends upon numerous factors such as the type of meal ordered at the centre the most, the type of raw materials available nearest to a centre, etc. This chapter will mainly focus on one factor, that is, the type of meal ordered from a particular centre. The results will determine the demand for raw materials for the centre-meal combination [2, 3].

Many forecasting models have been developed in recent years, and almost all of them share the same basic idea, but follow paradigms from different fields [1]. But the present chapter aims at the performance of the CatBoost Regressor. In the next section, the theoretical structure will be discussed in order to support the model proposed in this study.

2 Literature Survey

Online food ordering is the way of ordering food through an Internet site or a mobile phone application, generally handled by a particular restaurant or a mediator service (such as Zomato, Swiggy, Dunzo, etc.). The product may either be a ready-to-eat food product (delivered directly from a licenced home kitchen, restaurant, or a ghost kitchen) or food not explicitly prepared for direct consumption (like vegetables directly from a farm, frozen meats, etc) [3, 4]. The history dates back to 1994, when the famous pizza chain, Pizza Hut started accepting orders online. Since then, restaurants, hotels, groceries, etc. are majorly accepting orders through online mode. In online food orders placed at restaurants/chains, the restaurants prefer to use their own website/ app, or opt to employ a delivery vendor. Usually, food delivery orders are on demand and meant to be consumed immediately, including hot meals already prepared. Ordering food for delivery usually includes contacting a local restaurant or chain online or by telephone. For deliveries in large cities, it is not possible to deliver the meal on time and fresh [5]. Thus, restaurants have their own chains/franchises in major landmarks, which helps them to attract customers from the whole city.

According to a 2019 market analysis on food delivery services, it was found that the global online ordered food delivery market was valued at ₹7.1755 trillion and is expected to rise approximately 9% a year, approaching ₹10.267 trillion by 2023 [5]. The study defined the market as follows:

1. Meals ordered online that the restaurants deliver directly, either ordered via a portal (e.g. Zomato) or through its website.
2. Online meal orders and deliveries all driven by a platform (like Swiggy).
3. Online orders for takeaway in the restaurant by the customer. Excluding orders made via phone call (Fig. 1).

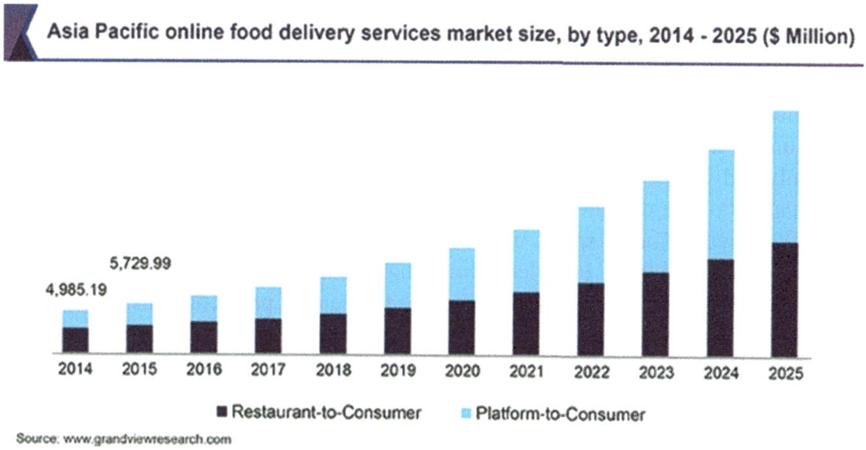

Fig. 1 Market analysis of online food delivery system [5]

3 Methodology

3.1 CatBoost Regressor

CatBoost is based on the Gradient Boosting library. Gradient boosting is an effective machine-learning technique that is extensively used in various types of business challenges like fraud detection, recommendations, demand/ supply forecast, and it achieves state-of-the-art results. Also, decision trees have been found as the base predictors when it comes to the implementation of gradient boosting algorithms. While dealing with pure numerical data, it is suitable to use decision trees, but during implementation many datasets contain categorical features that sometimes act as most important features for prediction [6].

A standard solution for dealing with categorical features that are not necessarily equivalent to one another is to substitute each feature with a number at the time of pre-processing [6, 7].

Gradient Boosting Algorithms like XGBoost, LightGBM, etc. deals with categorical data through one-hot encoding of such features before training the model. Unlike other algorithms, CatBoost does not require one-hot encoding of categorical features during model building.

CatBoost, similar to the various Gradient Boosting strategies, fabricates new trees to surmise the angle of the present tree. CatBoost utilizes a more efficient procedure that reduces overfitting and permits to utilize the entire dataset for preparation [6]. To be specific, a random permutation of the dataset in use is carried out, and an average label value for each example is computed. CatBoost comprises two boosting models, Ordered and Plain. Here, Ordered boosting is used [7].

In CatBoost, oblivious trees are used as base predictors. In such trees, the same criterion of splitting is used over a whole level of the tree point. Such trees are balanced and less likely to undergo overfitting [6]. CatBoost model evaluator makes use of these facts: It first converts all used one-hot encoded features, float features and statistics into binary and then uses those binary features to compute model predictions [6, 7].

Assuming a dataset having $\{(x_k, y_k)\}_{k=1}^{n}$, , where x_k is a vector of features chosen randomly and y_k is a target, initially CatBoost generates $s + 1$ independent random permutations of the dataset $(\sigma_0, \ldots, \sigma_s)$ [6, 7]. Now, permutations $\sigma_1, \ldots, \sigma_s$ will be used for evaluation of splits, while σ_0 will be used for choosing the leaf values of the trees obtained [7]. In ordered boosting, we have several supporting models $M_1, \ldots M_n$such that each M_i model will act as the current prediction for i the sample into the permutation σ_r [7]. In order to prevent "prediction shift", CatBoost makes use of permutations such that $\sigma_1 = \sigma_2$. This technique prevents the use of target for training of model for "Target Statistic" calculation or "gradient estimation".

Algorithm 1: Ordered Boosting

input: $\{(x_k, y_k)\}_{k=1}^{n}, I$;

$\sigma \leftarrow random\ permutation\ of\ [1, n]; M_i \leftarrow 0\ for\ i = 1..n;$

for $t \leftarrow 1$ **to** I **do**
  for $i \leftarrow 1$ **to** n **do**
    $r_i \leftarrow y_i - M_{\sigma(i)-1}(x_i);$
  for $i \leftarrow 1$ **to** n **do**
    $\Delta M \leftarrow LearnModel((x_j, r_j): \sigma(j) \leq i);$
    $M_i \leftarrow M_i + \Delta M;$
return M_n

Note: Originally published in "*CatBoost: unbiased boosting with categorical features support*" [4]

How CatBoost works?
1. Initially, one categorical feature is chosen to start with, let us say x.
2. Next, randomly a row is chosen from the training set, and a random level of this categorical feature is exchanged with a number.
3. The number chosen is generally based on the target variable depending on the category level. To put it another way, the target number is based on the estimated output variable.
4. A splitting function is used to construct two different sets of training data: One set with all the features will have a greater target variable, and the other set with smaller target variables.

4 Implementation

Here, Python3 is used to implement the model available in the CatBoost library by Yandex. To implement the CatBoost regressor, Online Food Delivery System's dataset is used (available online). The dataset contains historical data from 77 distinct delivery centres each of either Type A, Type B or Type C. The time frame of the data provided is 145 weeks. It contains several categorical features that have a direct impact on the prediction like operating area, type of cuisine, category of cuisine, etc. The aim is to perform regression analysis on the number of orders for the upcoming 10 weeks, which will tell about the required raw materials. Figure 2 is a glimpse of the dataset used in this study.

An exploratory analysis of the dataset was done and is shown below to observe the relationship between the variables (Figs. 3, 4, 5, 6, 7 and 8).

CatBoost offers a versatile interface for parameter selection and tuning. The parameters are selected after observing different values through grid search, which builds a model for every combination of hyperparameters specified and evaluates each model. The considered values for the parameters were: learning rate = [0.02, 0.03, 0.1]; tree depth = [4, 6, 8]; l2 regularization = [1, 3, 6, 9, 10]. The below table shows the best combinations where the resulting RMSE values were best (Table 1 and Fig. 9).

As RMSE (Root Mean Squared Error) is the loss function used in this analysis, the data needs to be converted into log values. Some of the features were even transformed for better understanding of the model. In this dataset, almost all the features were equally important for the prediction, but to achieve a better result, some extra features were also added like discount on base price, discount ratio, difference between price, certain *lag* features and *ewm* (exponentially weighted mean) features.

Below are shown some of the features participating in the forecast (Fig. 10).

Root Mean Square Error (RMSE) is the standard deviation of the residuals (prediction errors). It is the square root of MSE (Mean Squared Error) [8, 10]. In case of unbiased estimators, RMSE is just the square root of variance, which is actually Standard Deviation [2, 8]. In case of RMSLE (Root Mean Squared Logarithmic Error), the log of the predictions and actual values are taken. So basically, changes occur in the variance that is being measured [2].

Table 1 RMSE values for different combinations of hyperparameters

Learning rate	Tree depth	L2 regularization	**RMSE** (evaluation metric)
0.02	4	6	0.4547964942
0.03	6	6	0.443667502
0.1	6	9	0.4441142703
0.02	8	10	0.4396277843

	id	week	center_id	meal_id	checkout_price	base_price	\
0	1379560	1	55	1885	136.83	152.29	
1	1018704	2	55	1885	135.83	152.29	
2	1196273	3	55	1885	132.92	133.92	
3	1116527	4	55	1885	135.86	134.86	
4	1343872	5	55	1885	146.50	147.50	
5	1493612	6	55	1885	146.53	146.53	
6	1110832	7	55	1885	145.53	146.53	
7	1461167	8	55	1885	146.53	145.53	
8	1102364	9	55	1885	134.83	134.83	
9	1018130	10	55	1885	144.56	143.56	

	emailer_for_promotion	homepage_featured	num_orders	city_code_x	...	\
0	0	0	177	647	...	
1	0	0	323	647	...	
2	0	0	96	647	...	
3	0	0	163	647	...	
4	0	0	215	647	...	
5	0	0	285	647	...	
6	0	0	148	647	...	
7	0	0	135	647	...	
8	0	0	175	647	...	
9	0	0	175	647	...	

	center_type_y	op_area_y	category_y	cuisine_y	city_code	region_code	\
0	TYPE_C	2.0	Beverages	Thai	647	56	
1	TYPE_C	2.0	Beverages	Thai	647	56	
2	TYPE_C	2.0	Beverages	Thai	647	56	
3	TYPE_C	2.0	Beverages	Thai	647	56	
4	TYPE_C	2.0	Beverages	Thai	647	56	
5	TYPE_C	2.0	Beverages	Thai	647	56	
6	TYPE_C	2.0	Beverages	Thai	647	56	
7	TYPE_C	2.0	Beverages	Thai	647	56	
8	TYPE_C	2.0	Beverages	Thai	647	56	
9	TYPE_C	2.0	Beverages	Thai	647	56	

	center_type	op_area	category	cuisine
0	TYPE_C	2.0	Beverages	Thai
1	TYPE_C	2.0	Beverages	Thai
2	TYPE_C	2.0	Beverages	Thai
3	TYPE_C	2.0	Beverages	Thai
4	TYPE_C	2.0	Beverages	Thai
5	TYPE_C	2.0	Beverages	Thai
6	TYPE_C	2.0	Beverages	Thai
7	TYPE_C	2.0	Beverages	Thai
8	TYPE_C	2.0	Beverages	Thai
9	TYPE_C	2.0	Beverages	Thai

Fig. 2 Training dataset (first 10 rows)

$$RMSE = \sqrt{\frac{1}{n}\sum_{i=0}^{n}\left(y_i - \hat{y}_i\right)^2} \tag{1}$$

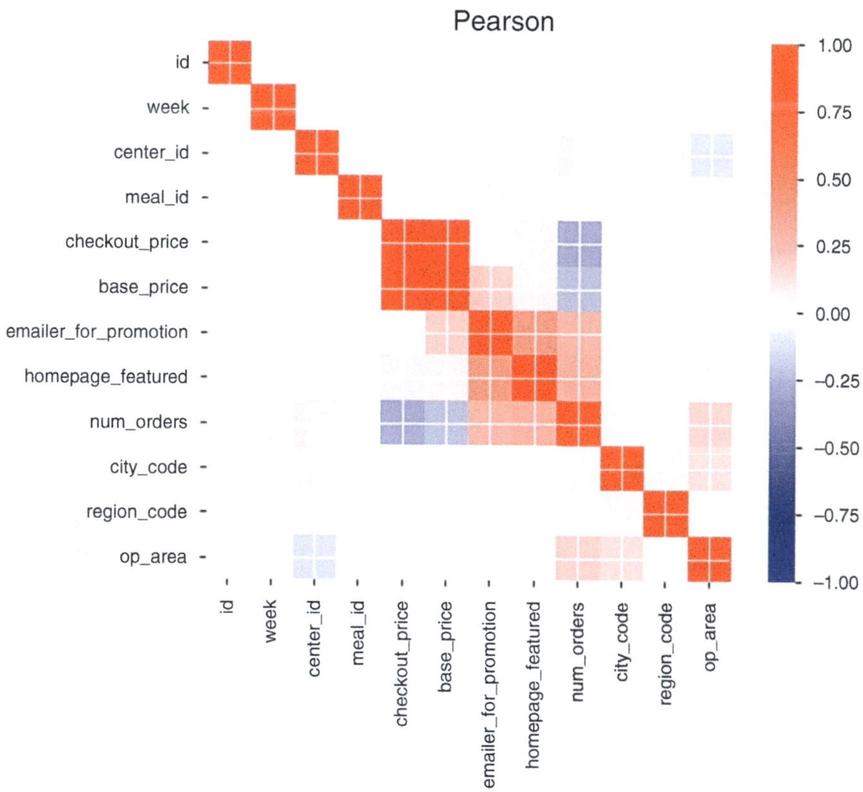

Fig. 3 Pearson correlation analysis between variables

$$RMSLE = \sqrt{\frac{1}{n}\sum\nolimits_{i=0}^{n}(\log\ (p_i + 1) - \log(a_i + 1))^2} \quad (2)$$

Once the features are decided, the data is split into train, validation and test. But here, the train data is divided for validation using K-Fold Cross Validation. And finally, the regression model is applied to the test dataset resulting in the number of orders for the next 10 weeks data.

5 Analyses and Results

This research on the demand forecasting of an Online Food Delivery service has brought some insightful points to notice. The evaluation metric, Root Mean Squared Logarithmic Error value obtained in this study was 0.0887.

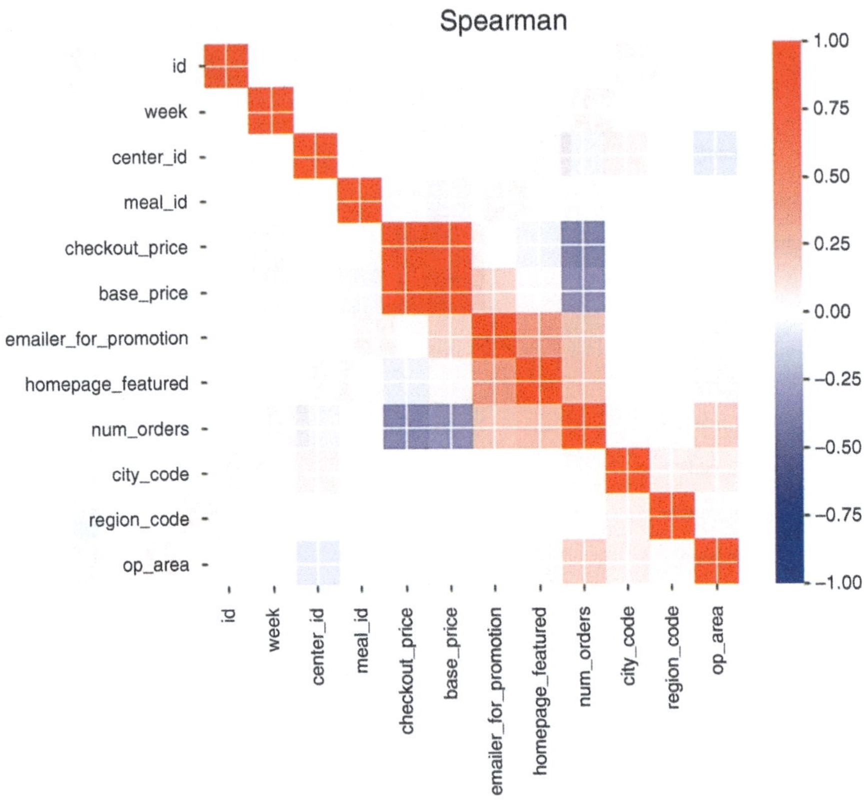

Fig. 4 Spearman correlation analysis between variables

When compared to other studies, the following results we obtained (Table 2).

The fit of the model can be observed in the below figure (Fig. 11), which depicts the sensitivity analysis of the model.

For a precise output of a forecasting model, adjustments must be made to the parameters for the technique applied. CatBoost, just like other machine learning packages, provides with a way to compare the feature (variable) importance [6]. This can help in building a better model, and allows for easier interpretation of results. Figure 12 is a plot showing the Feature Importance.

It is observed that avg_orders (average orders) have the highest importance among all variables. Evidently, CatBoost has shown some state-of-the-art results with most of the popular datasets containing categorical features. In this chapter, it shows more than expected results.

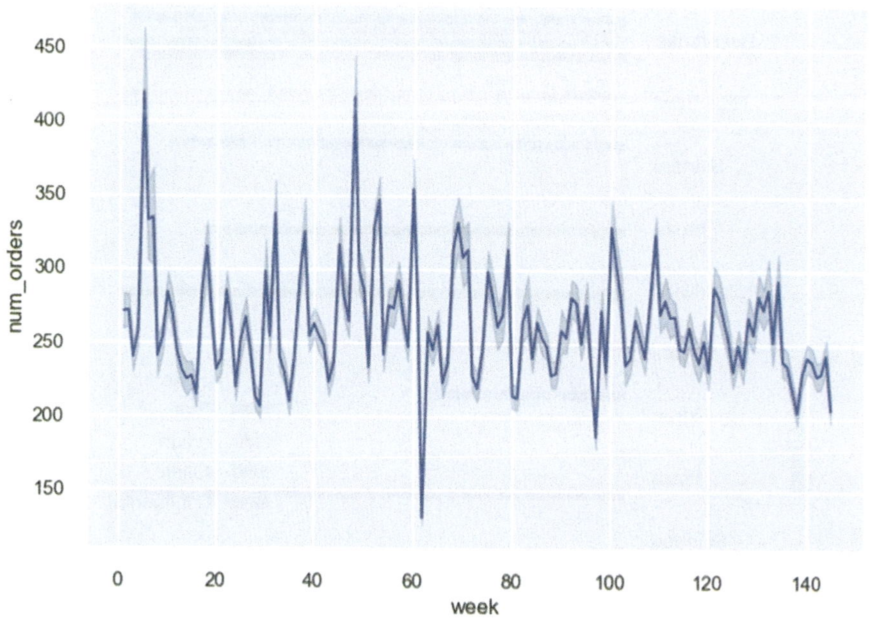

Fig. 5 Weeks vs number of orders

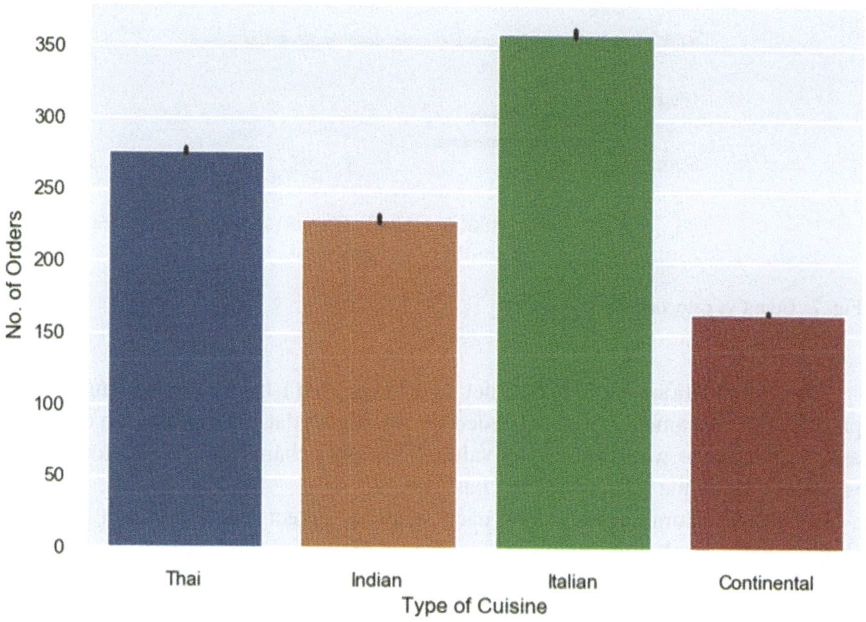

Fig. 6 Types of cuisine vs number of orders

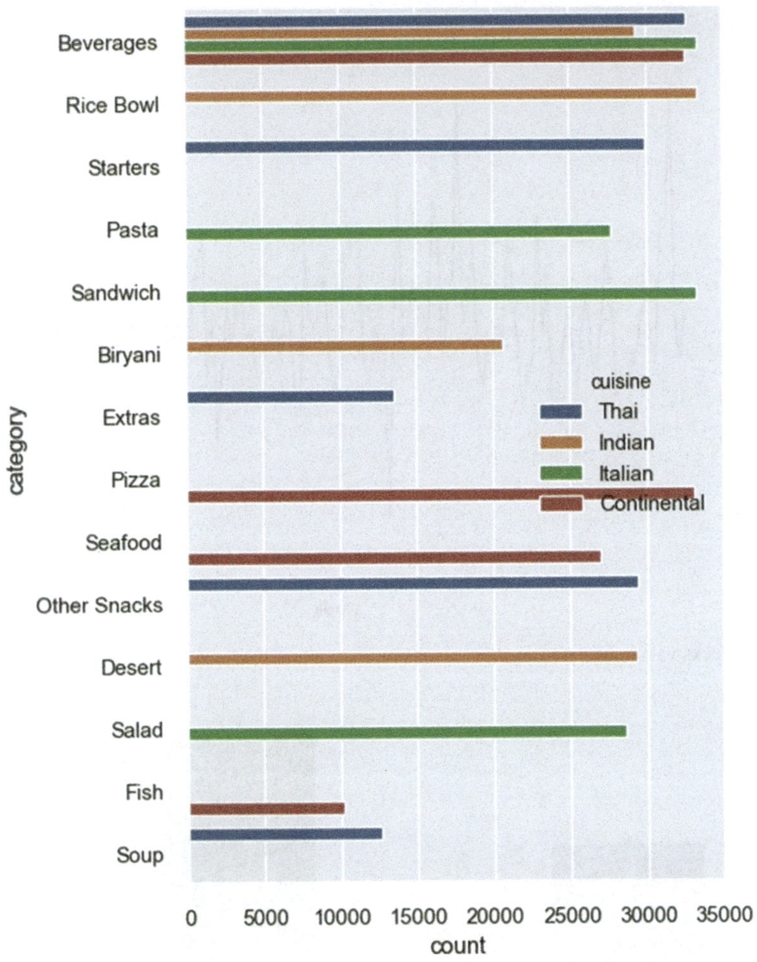

Fig. 7 Count vs categories

Also, applications for the Internet of Things (IoT) have evolved immensely, producing a vast amount of data needed for intelligent data processing. To optimize asset utilization as well as customer value, IoT supply chain management tools mine valuable data inside and outside the business [9].

IoT devices communicate with each other to collect and share data [9]. The results obtained through the study can provide insightful results for the data collected through the IoT devices and help in forecasting the demand of raw materials as well as automate the overall process of supply chain management (Fig. 13).

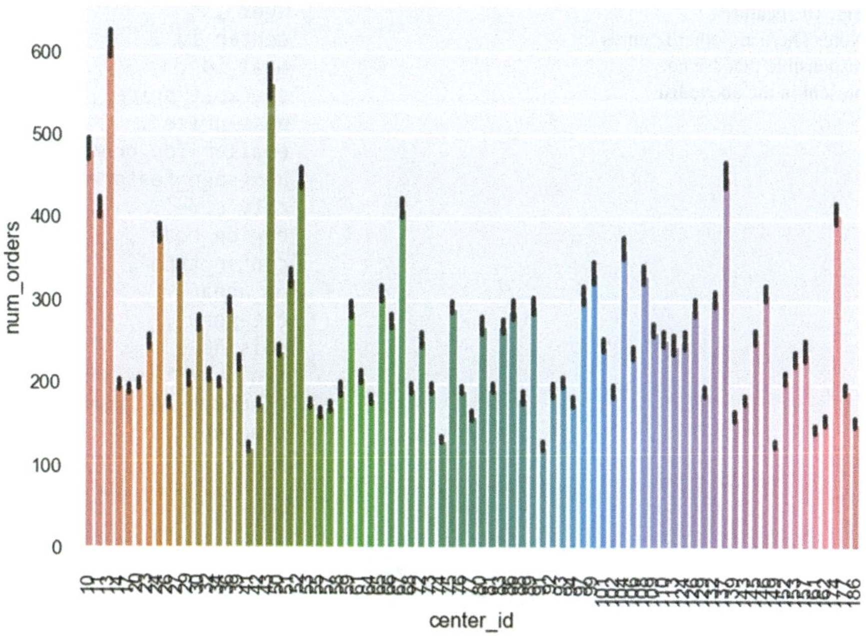

Fig. 8 Centre ID vs number of orders

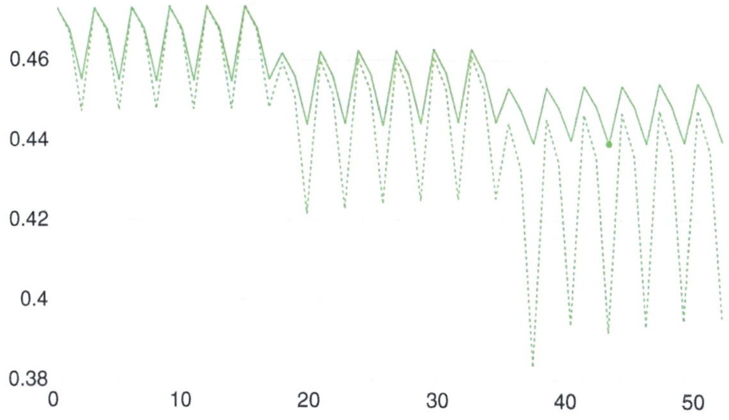

Fig. 9 Plot showing the variation in RMSE value based on the hyperparameter values

Fig. 10 Features
Note: There are other features responsible that are not present in the above list

```
['week',
 'center_id',
 'meal_id',
 'checkout_price',
 'base_price',
 'emailer_for_promotion',
 'homepage_featured',
 'city_code',
 'region_code',
 'center_type',
 'op_area',
 'category',
 'cuisine',
 'train_or_test',
 'discount_on_base',
 'discount_ratio',
 'price_last_curr_diff']
```

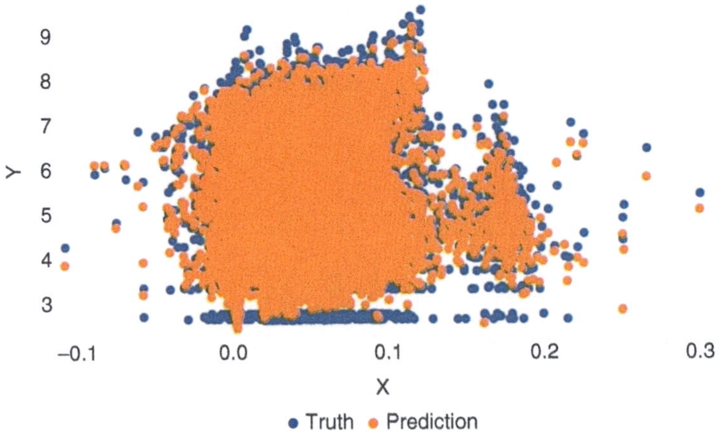

Fig. 11 True values vs predicted values

Table 2 Performance of different models on the food demand dataset

	LightGBM	XGBoost	**CatBoost**
RMSE	0.583	0.813	**0.446**
RMSLE	0.218	0.146	**0.0887**

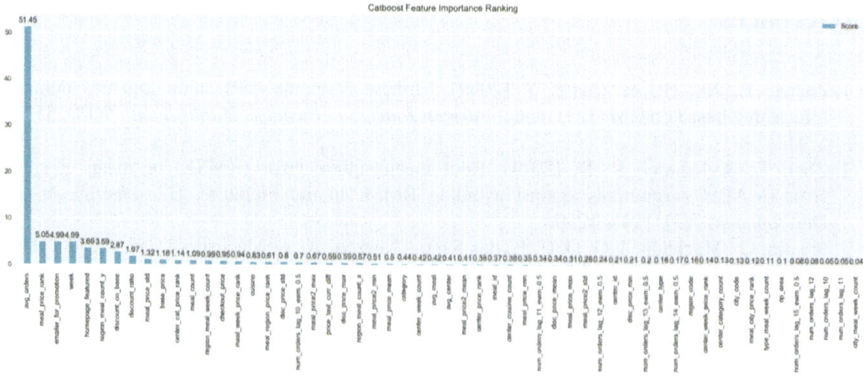

Fig. 12 Feature importance Score

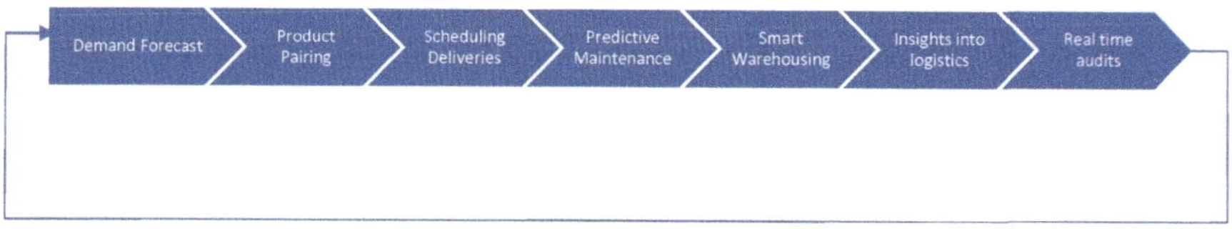

Fig. 13 How IoT enhances efficiency in the food supply chain

6 Conclusion

From the results obtained from this analysis, the following conclusions can be drawn:

1. The CatBoost model applies shift and ewm (exponentially weighted mean) features to the dataset as the data used contains historical data values.
2. CatBoost proves to offer improved performance as compared to other regression models that also deals with categorical data.
3. The historical data has a great and major impact on predicting the future values.
4. For any retailer, manufacturer or supplier, forecasting product demand is essential for its business' growth. Demand forecasting helps determine the quantities that should be purchased, produced and shipped.

In general, accurate demand forecasts help the manufacturers/suppliers obtain a correct idea of the raw materials requirements in future, which leads to high levels of customer services. On the other hand, erroneous forecasts can lead to high-cost operations and a poor customer service.

References

1. Zhang, Y., Xu, H., & Zheng, Y. (2009). Chinese residents cold chain logistics demand forecasting based on GM (1, 1) model. *African Journal of Business Management, 6*(14), 5136–5141.
2. Chai, T., & Draxler, R. R. (2014). Root mean square error (RMSE) or mean absolute error (MAE)? –arguments against avoiding RMSE in the literature. *Geoscientific Model Development, 7*(3), 1247–1250.
3. Pigatto, G., Machado, J. G. D. C. F., dos Santos Negreti, A., & Machado, L. M. (2017). Have you chosen your request? Analysis of online food delivery companies in Brazil. *British Food Journal, 119*, 639.
4. Keeble, M., Adams, J., Sacks, G., Vanderlee, L., White, C. M., Hammond, D., & Burgoine, T. (2020). Use of online food delivery services to order food prepared away-from-home and associated sociodemographic characteristics: A cross-sectional, multi-country analysis. *International Journal of Environmental Research and Public Health, 17*(14), 5190.
5. Zulkarnain, K., Ahasanul, H., & Selim, A. (2015). Key success factors of online food ordering services: An empirical study. *Malaysian Institute of Management, 50*(2), 19–36.
6. Dorogush, A.V., Ershov, V. & Gulin, A. (2018). CatBoost: gradient boosting with categorical features support. arXiv preprint arXiv:1810.11363.
7. Prokhorenkova, L., Gusev, G., Vorobev, A., Dorogush, A. V., & Gulin, A. (2018). CatBoost: Unbiased boosting with categorical features. In *Advances in neural information processing systems* (pp. 6638–6648).
8. Hastie, T., Tibshirani, R., & Friedman, J. (2009). *The elements of statistical learning: Data mining, inference, and prediction*. Springer Science & Business Media.
9. Adi, E., Anwar, A., Baig, Z., & Zeadally, S. (2020). Machine learning and data analytics for the IoT. In *Neural computing and applications* (pp. 1–29).
10. Da Veiga, C. P., Da Veiga, C. R. P., Catapan, A., Tortato, U., & Da Silva, W. V. (2014). Demand forecasting in food retail: A comparison between the Holt-winters and ARIMA models. *WSEAS Transactions on Business and Economics, 11*(1), 608–614.

GLCM Feature-Based Texture Image Classification Using Support Vector Machine

R. Anand, T. Shanthi, R. S. Sabeenian, M. E. Paramasivam, and K. Manju

Abstract Texture is one of the important characteristics of an image, and its features can be used to identify specific region of interest from the image. Texture features can be extracted for any kind of images like RGB, monochrome, aerial, satellite images. This chapter describes an approach for texture image classification based on Gray-level co-occurrence matrix (GLCM) features and machine learning algorithm like support vector machine. Ten different GLCM features were extracted and fed as input to support vector machine for classification. The proposed method is trained and tested with dataset collected from center for image analysis, Swedish University. In recent days, machine learning (ML) methods are highly used to mimic the complex mathematical expressions of texture features. The main objective of this chapter is to demonstrate the state of the art of ML models in texture image prediction and to give insight into the most suitable models.

Keywords Support vector machine · Texture image · Machine learning · Confusion Matrix · Gray-level co-occurrence matrix

1 Introduction

Textures are complex visual patterns, composed of spatially organized entities that have characteristic brightness, color, shape, and size. Texture is a description of the spatial arrangement of color or intensities in an image or selected region of an image. Image texture is a function of spatial variation of pixel intensities. Texture is homogenous for human visual system. Some texture descriptions are mainly used in many real-time applications; these features are contrast, brightness, likeness, regularity, roughness etc., Texture involves spatial distribution of gray-level image,

R. Anand (✉) · T. Shanthi · R. S. Sabeenian · M. E. Paramasivam · K. Manju
Department of Electronics and Communication Engineering, Sona College of Technology, Salem, Tamil Nadu, India
e-mail: anand.r@sonatech.ac.in

A. Haldorai et al. (eds.), *3rd EAI International Conference on Big Data Innovation for Sustainable Cognitive Computing*, EAI/Springer Innovations in Communication and Computing, https://doi.org/10.1007/978-3-030-78750-9_10

different levels of resolutions, usually extracted form single band. Texture features from different bands (multiple bands) of the image are generally different and have capability of discriminating different land use and land cover types. Image segmentation using texture features is partitioning an image into regions that have homogenous properties with respect to texture. Image classification using texture features is assigning a label to an unknown sample image or a portion of image with a known texture class. Texture synthesis is building a model of image texture, which can be used for generating the texture.

2 Literature Survey

2.1 Statistical Approaches

Images with varying textures have certain characteristics that can be extracted statistically. Generally, statistical approaches have four prevalent methods like Gray-Level Co-occurrence Matrix (GLCM), Histogram Method, Auto-Correlation Method, and Morphological operation. These four methods have both advantages and disadvantages. Out of these four methods, this chapter adopts Gray-level co-occurrence matrix (GLCM) from which ten different features are extracted. These features are elaborated in the next section, and the literature on statistical approaches is shown in Table 1.

Table 1 Strength and weakness of statistical approach methods for texture image classifications

Methods	Strengths	Weakness
Gray-level co-occurrence Matrix (GLCM) [8]	• Spatial relationship of pixels with ten different statistical computations. • (contrast, energy, homogeneity, mean, standard deviation, entropy, RMS, variance, smoothness, IDM) • Accuracy rate will be high.	• High computational time. • Optimum movement vector is problematic. • It requires feature selection procedure. • Accuracy depends on the offset rotation.
Histogram Method [9]	• Less computations. • Invariant to conversion and rotation. • Mathematically solvable.	• Sensitive to noise. • Low recognition rate.
Auto-correlation Method [7]	• It overcomes illumination distortion and it is robust to noise. • Low computational complexity.	• Real-time applications for large images need high computations. • Not suitable for all kinds of textures.
Morphological operation [6]	• Good efficient aperiodic image texture.	• Morphological operation cannot be applicable for periodic images.

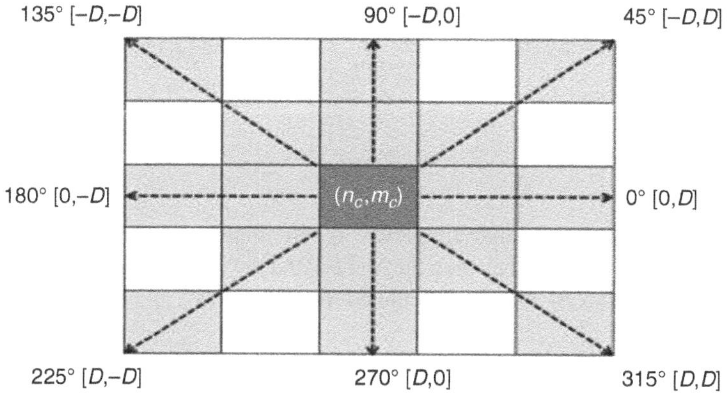

Fig. 1 Co-occurrence matrix

2.2 GLCM

The Gray-Level Co-occurrence Matrix method—likewise often named as the Spatial Gray-Level Dependence Matrix (SGLDM) Method. This method is based on the intensity values of the image. The features when fed to the machine learning algorithm perform the classification effectively [1, 2].

The second-order statistics are attained by considering a set of pixels related to each other in positive three dimensions. The Gray-Level Co-occurrence matrices provide rare mathematical statistics on the texture. Texture feature calculation uses the content of the GLCM to give a measure of the variation in intensity at the pixel of interest. GLCM matrix is a square matrix that has the same number of rows and same number of columns. GLCM matrix is N × N matrix, where N denotes the number of possible gray levels in an image. For example, a three-bit image will have a GLCM matrix of size 8 × 8. The rows and columns are the gray values (0–7). GLCM matrix of image depends on the direction and offset values. The direction can be anyone among the eight possible directions as shown in Fig. 1. The offset represents distance between pixels.

This chapter utilizes ten different features that are computed from GLCM Matrix [13]. These ten features are elaborated as follows.

2.2.1 Entropy

Entropy measures the disorder or complexity of an image. If the entropy is large when the image is not texturally uniform and if the texture is complex, then entropy becomes high. Energy and entropy are inversely proportional to each other; it means if entropy is high, energy will be low and vice versa. Entropy can be calculated for image $f(x, y)$ using the following equation:

$$\text{Entropy (ENT)} = \sum_{x=0}^{N-1}\sum_{y=0}^{N-1} f(x, y) * \log(f(x, y))$$

2.2.2 Contrast

Contrast measures the spatial frequency of an image and different moment of GLCM. It is the difference between the highest and the lowest values of a contiguous set of pixels. Contrast can be calculated for image $f(x, y)$ using the following equation:

$$\text{Contrast} = \sum_{x=0}^{N-1}\sum_{y=0}^{N-1} (x - y)^2 f_{xy}$$

2.2.3 Mean

Mean is the normal value of the whole amount of data. Mathematically,

$$\text{Mean }(\mu i) = \frac{\sum_{x=0}^{N-1}\sum_{y=0}^{N-1} f(x, y)}{MxN}$$

2.2.4 Standard Deviation

The standard deviation is the square root of variance divided by the entire. number of samples. Mathematically, it is expressed as

$$SD = \sqrt{\frac{\sum |x - \mu|^2}{N}}$$

2.2.5 Variance

This statistic measures the heterogeneity, and it is strongly correlated with first-order statistical variable such as standard deviation. Variance can be calculated for image $f(x, y)$ using the following equation:

$$\text{Variance} = \sum_{x=0}^{N-1}\sum_{y=0}^{N-1} (i - \mu)^2 f_{xy}$$

where 'μ' is the mean of an input image.

2.2.6 Homogeneity

Homogeneity is the consistency in the arrangement of an input image $f(x, y)$. If the arrangement of an input image follows a regular pattern, then it is said to be homogenous. Mathematically, it can be expressed by the following equation:

$$\text{Homogenous } f(x, y) = \sum_{x=0}^{N-1} \sum_{y=0}^{N-1} \frac{f_{xy}}{1 + |x - y|^2})$$

2.2.7 RMS Contrast

Root Mean Square (RMS) measures the standard deviation of the pixel intensities. It does not depend upon any angular frequency or spatial distribution of contrast of an input image. Mathematically, it can be expressed as

$$\text{RMS Contrast} = \sqrt{\frac{1}{MN} \sum_{x=0}^{N-1} \sum_{y=0}^{N-1} (I_{ij} - \bar{I})^2}$$

where 'I' is the normalized pixel intensity values between 0 and 1.

2.2.8 Smoothness

If the input image $f(x, y)$ is highly correlated between the adjacent pixels, then we can say the image is autocorrelated (the autocorrelation of input data with itself after shifting one pixel). The autocorrection of input image can be calculated by the following equation:

$$\text{autocorrelation } f(x, y) = \iint f(x, y) * \left(f(x + \overline{x})(y + \overline{y})\, dx\, dy\right.$$

2.2.9 Inverse Difference Moment (IDM)

IDM measures the local similarity between of an image. It has the feature of measuring the closeness and distribution of the GLCM elements with respect to GLCM diagonal. It also has the property to determine the texture of the image. The frequency of co-occurrence of set of pixels is improved when their values of gray level are close enough, and this increases the IDM value.

$$\text{Inverse Difference Moment (IDM)} = \sum_{x=0}^{N-1} \sum_{y=0}^{N-1} \frac{f_{xy}}{1 + |x - y|^2}$$

3 Proposed Method

Texture features of an image are calculated considering only one band at a time. Channel information can be consolidated using PCA before calculating texture features. Texture features of an image can be used for both supervised and unsupervised image classification. In this chapter, we have used support vector machine (SVM) for texture image classification. This method falls under the category of supervised machine learning [10]. Support vector machine was introduced in 1992 as a supervised machine leaning algorithm. This algorithm gained its popularity because of its higher accuracy rate and minimum error rate. SVM is one of the best example for "Kernel Method' which is the key area of machine learning that is shown in Fig. 2. The idea behind SVM is to make use of nonlinear mapping function φ, that transforms data in input space to data in feature space in such way that it becomes a linearly separable [3]. The SVMs then automatically discover the optimal separating hyperplane, which is nothing but a complex decision surface. The equation of hyperplane is derived from a line equation $y = ax + b$. However, even though hyperplane is a line, its equation is shown below [11, 12], where "w and x" are the vectors, and it can be computed by dot matrix of these two vectors [5].

$$w\hat{}T\ x = 0$$

Any hyperplane can be written as the set of points, x satisfying the equation $w.\,x + b = 0$. A optimum hyperplane separates the data points with minimum loss function. Two such hyperplanes are chosen and based on the values obtained; they are classified as class 1 and class 2 as given below.

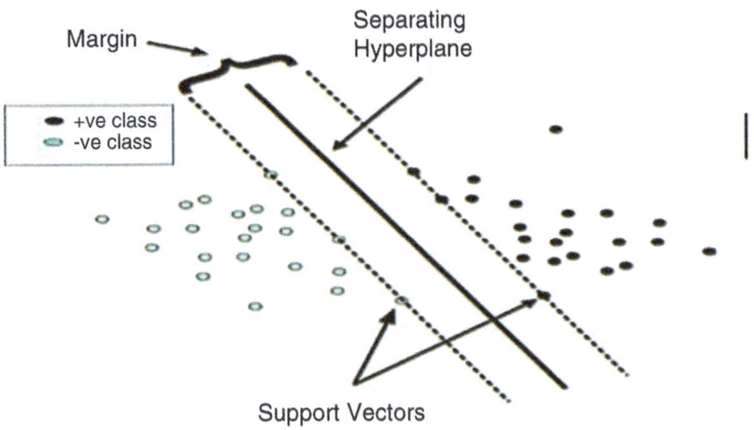

Fig. 2 Separating hyperplane in SVM

$$w.x + b \geq 1 \text{ for } x\_i \text{ having class 1}$$

$$w.x + b \leq -1 \text{ for } x\_i \text{ having class 2}$$

Here, optimizing problem may occur because the goal is to maximize the margin among all the possible hyperplanes meeting the constraints. The hyperplane with the small ||w|| is chosen because of the biggest margin it provides. The pseudo codes for optimizing problem are given below:

$$\text{Minimize } in\,(w, b)\,,\ \|w\|\,,\ \text{subject } to\ y\_i\ \ (w.x + b) \geq 1$$

The solution to the above equation is computing the values of (w, b),with minimum possible margin. The equation that satisfies the constraints will be considered as the equation of the optimal hyperplane [14, 15].

3.1 Dataset Description

The dataset from center for image analysis in Swedish university of agriculture sciences [4], Uppsala University has been used in this chapter. Totally, 4480 images of 28 different texture classes were taken using Canon EOS 550d DSLR camera, which is shown in Fig. 3. Each texture class has around 160 images, in that 112 images are used for training and 48 images are used for testing which is shown in table [2]. Fig. 4 shows the complete flow chart of the proposed method for texture image classifications using GLCM features. First step, the segmented image is resized to [576 x 576] (Table 2).

The performance of the proposed system is measured in terms of sensitivity, specificity, accuracy, precision, false positive rate and false negative rate, which is shown in Fig. 5. The sensitivity and specificity are important measures in classification. The accuracy of the system represents the exactness of the system with respect to classification. The precision indicates the ability of the system to produce the same result when tested at different measurements. The overall classification accuracy of the proposed system is around 99.4% with the precision of 92.4% . The false negative rate and false positive rate is very minimum in the range of 0.003 and 0.085, respectively. The sensitivity and specificity of the system is around 91.5% and 99.7%.The texture classes 2,4,5,7,9,12, and 19 have been classified with good accuracy and precision when compared to other classes.

The GUI of the proposed system is shown in Fig. 6 (Table 3).

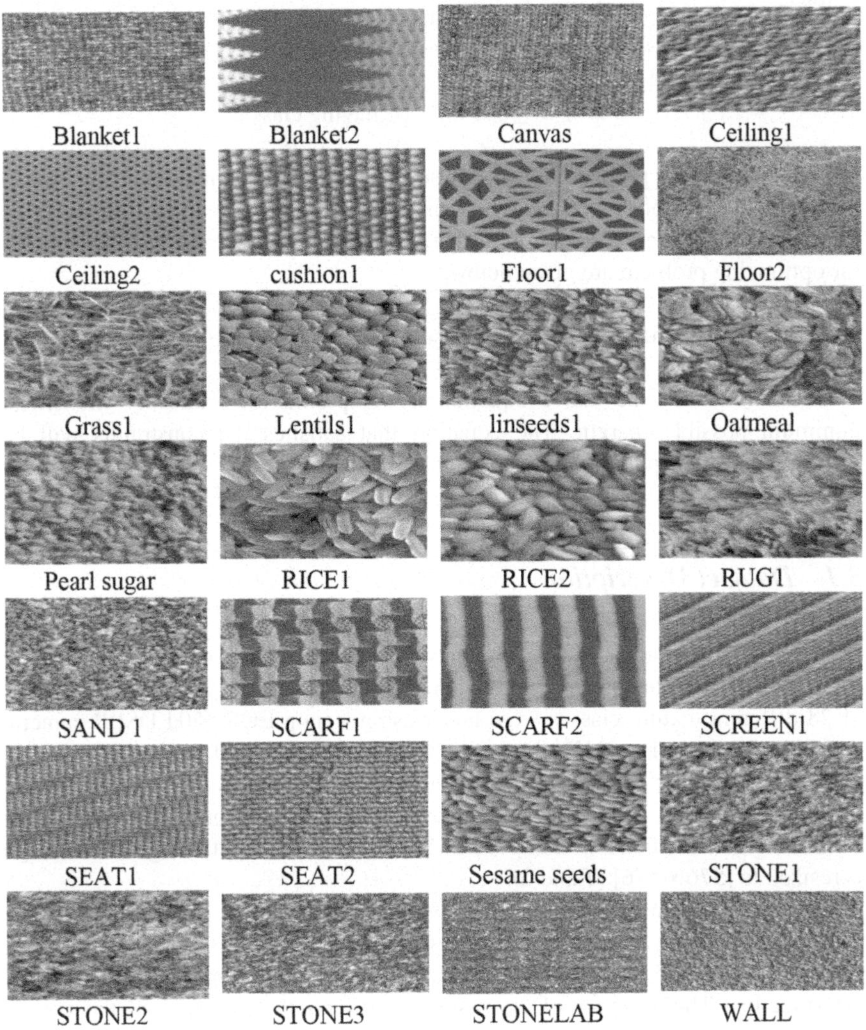

Fig. 3 Sample images from 28 different texture images

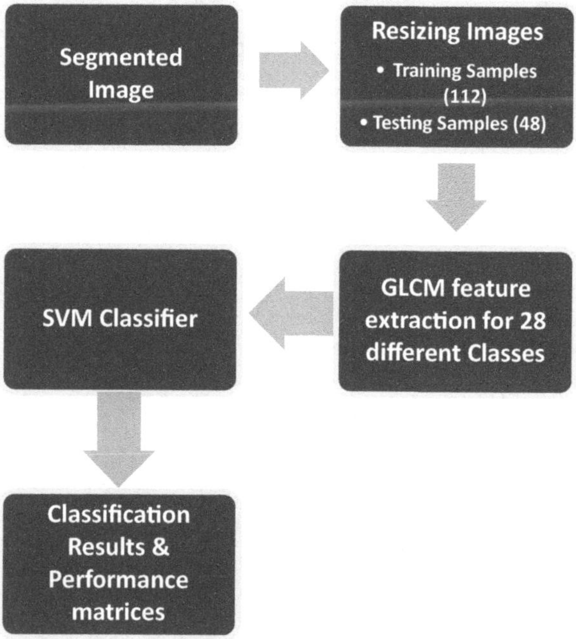

Fig. 4 Flow chart for produced texture image classifications

Table 2 Dataset descriptions

Class label	Training samples	Testing samples	Total samples
Blanket1	112	48	160
Blanket2	112	48	160
Canvas	112	48	160
Ceiling1	112	48	160
Ceiling2	112	48	160
cushion1	112	48	160
Floor1	112	48	160
Floor2	112	48	160
Grass1	112	48	160
Lentils1	112	48	160
linseeds1	112	48	160
Oatmeal	112	48	160
Pearl sugar	112	48	160
RICE1	112	48	160
RICE2	112	48	160

(continued)

Table 2 (continued)

Class label	Training samples	Testing samples	Total samples
RUG1	112	48	160
SAND 1	112	48	160
SCARF1	112	48	160
SCARF2	112	48	160
SCREEN1	112	48	160
SEAT1	112	48	160
SEAT2	112	48	160
Sesame seeds	112	48	160
STONE1	112	48	160
STONE2	112	48	160
STONE3	112	48	160
STONELAB	112	48	160
WALL	112	48	160

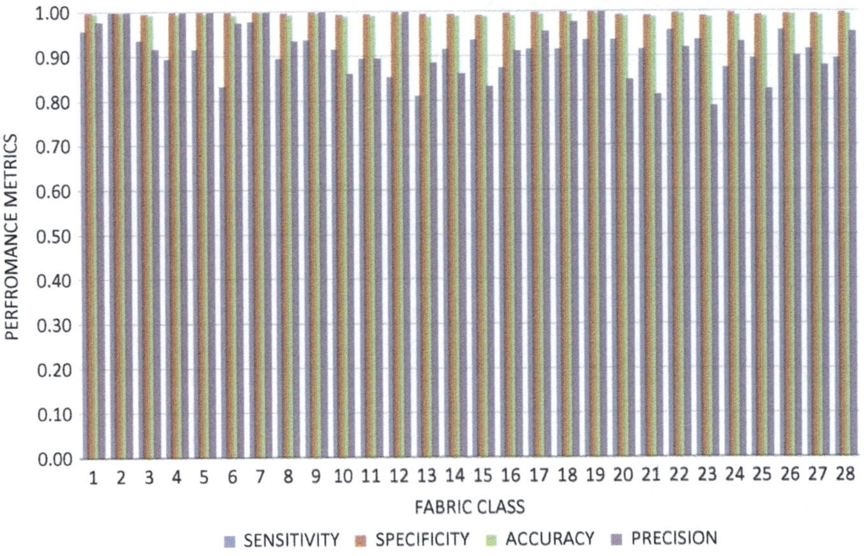

Fig. 5 Prediction using support vector machine

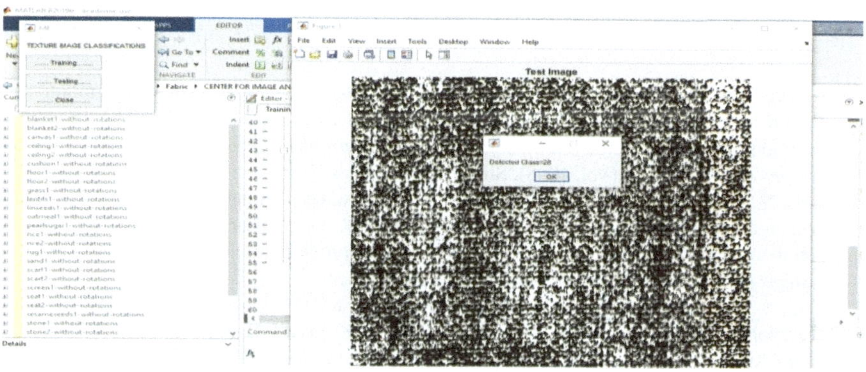

Fig. 6 Performance metrics evaluation for texture image classification

Table 3 Confusion matrix for 28 classes

Predicted class / Actual Class	1	2	3	4	5	6	7	8	9	10	11	12	13	14	15	16	17	18	19	20	21	22	23	24	25	26	27	28
1	46	0	1	0	0	0	0	0	0	0	0	0	0	0	0	0	0	0	0	0	0	0	0	0	0	1	0	0
2	0	48	0	0	0	0	0	0	0	0	0	0	0	0	0	0	0	0	0	0	0	0	0	0	0	0	0	0
3	0	0	45	0	0	0	0	2	0	0	0	0	0	0	0	0	0	0	0	0	0	0	0	0	0	0	1	0
4	1	0	0	43	0	0	0	0	0	0	0	0	1	0	0	0	0	0	0	0	0	0	0	0	0	1	2	0
5	0	0	0	0	44	0	0	0	0	0	0	0	0	0	0	0	0	0	0	3	0	0	0	0	0	0	1	0
6	0	0	0	0	0	40	0	0	0	0	0	0	0	3	1	0	0	0	0	0	0	0	4	0	0	0	0	0
7	0	0	0	0	0	0	47	1	0	0	0	0	0	0	0	0	0	0	0	0	0	0	0	0	0	0	0	0
8	0	0	0	0	0	1	0	43	0	0	0	0	2	0	0	0	0	0	0	0	0	0	0	1	1	0	0	0
9	0	0	0	0	0	0	0	0	45	0	0	0	0	0	0	2	0	0	0	0	0	0	0	1	0	0	0	0
10	0	0	0	0	0	0	0	0	0	44	0	0	0	2	2	0	0	0	0	0	0	0	0	0	0	0	0	0
11	0	0	0	0	0	0	0	0	0	0	43	0	0	2	1	0	0	0	0	0	0	0	2	0	0	0	0	0
12	0	0	0	0	0	0	0	0	0	4	0	41	0	0	0	0	0	0	0	0	0	0	3	0	0	0	0	0
13	0	0	0	0	0	0	0	0	0	0	3	0	39	0	0	0	2	0	0	0	0	0	3	0	0	0	0	0
14	0	0	0	0	0	0	0	0	0	2	0	0	0	44	2	0	0	0	0	0	0	0	0	0	0	0	0	0
15	0	0	0	0	0	0	0	0	0	1	0	0	0	0	45	2	0	0	0	0	0	0	0	0	0	0	0	0
16	0	0	0	0	0	0	0	0	4	0	0	0	0	0	0	42	0	0	0	0	0	0	0	0	2	0	0	0
17	0	0	0	0	0	0	0	0	0	0	0	0	0	0	0	0	44	0	0	0	3	1	0	0	0	0	0	0
18	0	0	0	0	0	0	0	0	0	0	2	0	0	0	0	0	0	44	0	0	2	0	0	0	0	0	0	0
19	0	0	0	0	0	0	0	0	0	0	0	0	0	0	0	0	0	1	45	2	0	0	0	0	0	0	0	0
20	0	0	0	0	0	0	0	0	0	0	0	0	0	0	0	0	0	0	0	45	3	0	0	0	0	0	0	0
21	0	0	0	0	0	0	0	0	0	0	0	0	0	0	0	0	0	0	0	3	44	1	0	0	0	0	0	0
22	0	0	0	0	0	0	0	0	0	0	0	0	0	0	0	0	0	0	0	0	2	46	0	0	0	0	0	0
23	0	0	0	0	0	0	0	0	0	0	0	0	0	0	3	0	0	0	0	0	0	0	45	0	0	0	0	0
24	0	0	0	0	0	0	0	0	0	0	0	0	2	0	0	0	0	0	0	0	0	0	0	42	4	0	0	0
25	0	0	0	0	0	0	0	0	0	0	0	0	0	0	0	0	0	0	0	0	0	0	0	1	43	3	0	0
26	0	0	0	0	0	0	0	0	0	0	0	0	0	0	0	0	0	0	0	0	0	0	0	0	2	46	0	0
27	0	0	0	0	0	0	0	0	0	0	0	0	0	0	0	0	0	0	0	0	0	2	0	0	0	0	44	2
28	0	0	3	0	0	0	0	0	0	0	0	0	0	0	0	0	0	0	0	0	0	0	0	0	0	0	2	43

4 Conclusion

Texture of an image is a description of the spatial arrangement of color or intensities in the image. Texture of image can be used to categorize the image into several classes. Ample number of texture features can be computed mathematically and used for image analysis. The method proposed in this chapter combines the texture features computed from GLCM matrix with that of the standard machine learning algorithm for image classification. The overall classification accuracy of the proposed system is around 99.4% with the precision of 92.4%. The classification accuracy of the system can be further improved by increasing the size of the dataset.

References

1. Anand, R., Shanthi, T., Nithish, M. S., & Lakshman, S. (2020). Face recognition and classification using GoogleNET architecture. In K. Das, J. Bansal, K. Deep, A. Nagar, P. Pathipooranam, & R. Naidu (Eds.), *Soft computing for problem solving. Advances in intelligent systems and computing* (Vol. 1048). Singapore: Springer.
2. Veni, S., Anand, R., & Vivek, D. (2020). Driver assistance through geo-fencing, sign board detection and reporting using android smartphone. In K. Das, J. Bansal, K. Deep, A. Nagar, P. Pathipooranam, & R. Naidu (Eds.), *Soft computing for problem solving. Advances in intelligent systems and computing* (Vol. 1057). Singapore: Springer.
3. Veni, S., Anand, R., & Vivek, D. (2020). Driver assistance through geo-fencing, sign board detection and reporting using android smartphone. In K. Das, J. Bansal, K. Deep, A. Nagar, P. Pathipooranam, & R. Naidu (Eds.), *Soft computing for problem solving. Advances in intelligent systems and computing* (Vol. 1057). Singapore: Springer.
4. G. Kylberg. The Kylberg texture dataset v. 1.0, Centre for Image Analysis, Swedish University of Agricultural Sciences and Uppsala University, External report (Blue series) No. 35. Available online at: http://www.cb.uu.se/gustaf/texture/
5. Anand, R., Veni, S., & Aravinth, J. (2016). An application of image processing techniques for detection of diseases on brinjal leaves using k-means clustering method. *2016 International Conference on Recent Trends in Information Technology* (ICRTIT). IEEE.
6. Sabeenian, R. S., & Palanisamy, V. (2009). Texture-based medical image classification of computed tomography images using MRCSF. *International Journal of Medical Engineering and Informatics, 1*(4), 459.
7. Sabeenian, R. S., & Palanisamy, V. (2008). Comparison of efficiency for texture image classification using MRMRF and GLCM techniques. *Published in International Journal of Computers Information Technology and Engineering (IJCITAE), 2*(2), 87–93.
8. Haralick, Robert, M., Shanmugam, K., & Dinstein, H. (1973). Textural features for image classification. *IEEE Transactions on systems, man, and cybernetics, 6*, 610–621.
9. Varma, M., & Zisserman, A. (2005). A statistical approach to texture classification from single images. *International Journal of Computer Vision, 62*(1–2), 61–81.
10. Suykens, Johan, A. K., & Vandewalle, J. (1999). Least squares support vector machine classifiers. *Neural Processing Letters, 9.3*, 293–300.
11. Scholkopf, B., & Smola, A. J. (2001). *Learning with kernels: Support vector machines, regularization, optimization, and beyond.* MIT press.
12. Bennett, K. P., & Demiriz, A. (1999). Semi-supervised support vector machines. In *Advances in neural information processing systems* (pp. 368-374).
13. Shanthi, T., Sabeenian, R. S., & Anand, R. (2020). Automatic diagnosis of skin diseases using convolution neural network. *Microprocessors and Microsystems, 103074.*
14. Shanthi, T., & Sabeenian, R. S. (2019). Modified Alexnet architecture for classification of diabetic retinopathy images. *Computers and Electrical Engineering, 76*, 56–64.
15. Hall-Beyer, M. (2000). GLCM texture: a tutorial. *National Council on Geographic Information and Analysis Remote Sensing Core Curriculum, 3.*

Improved Rider Optimization Algorithm-Based Link Aware Fault Detection (IROA-LAFD) Scheme for Securing Mobile Ad Hoc Networks (MANETs)

Sengathir Janakiraman, M. Deva Priya, G. Aishwaryalakshmi, T. Suganya, S. Sam Peter, S. Karthick, and A. Christy Jeba Malar

Abstract Securing communication in a dynamic network like Mobile Ad hoc Network (MANET) is considered as a crucial and demanding task. A number of diversified works are developed in the literature for mitigating harmful attacks that degrade the performance of the network. Faulty links need to be detected for attaining better performance in terms of reliability and availability of mobile nodes in the network. Moreover, attacks need to be detected through a single stage attack detection process in a proactive way for guaranteeing quality of service in the network. However, the detection accuracy of most of the existing approaches in the literature still possess room for improvement. In this paper, Improved Rider Optimization Algorithm-based Link Aware Fault Detection (IROA-LAFD) scheme is proposed for facilitating security by mitigating grey hole and black hole attacks with enhanced link stability. This IROA-LAFD scheme targets on efficient

---

S. Janakiraman
Department of Information Technology, CVR College of Engineering, Hyderabad, Telangana, India

M. Deva Priya (✉) · T. Suganya · S. Sam Peter
Department of Computer Science and Engineering, Sri Krishna College of Technology, Coimbatore, Tamil Nadu, India
e-mail: m.devapriya@skct.edu.in; suganya.t@skct.edu.in; sampeter.s@skct.edu.in

G. Aishwaryalakshmi
PSG Polytechnic College, Coimbatore, Tamil Nadu, India

S. Karthick
Department of Computer Science and Engineering, SRM Institute of Science and Technology, Ghaziabad, India
e-mail: karthics1@srmist.edu.in

A. Christy Jeba Malar
Department of Information Technology, Sri Krishna College of Engineering and Technology, Coimbatore, Tamil Nadu, India
e-mail: a.christyjebamalar@skct.edu.in

A. Haldorai et al. (eds.), *3rd EAI International Conference on Big Data Innovation for Sustainable Cognitive Computing*, EAI/Springer Innovations in Communication and Computing, https://doi.org/10.1007/978-3-030-78750-9_11

mitigation of packet dropping based on the steps that include the discovery of neighbour and route, detection of attack, analysis of links, transmission of secure packets and link fault detection using IROA Algorithm. In particular, IROA is used for exploring the link paths established among the trust nodes in the network. The simulation experiments of the proposed IROA-LAFD scheme is identified to increase the Packet Delivery Ratio (PDR) by 13.42%, throughput by 12.84% and the link fault detection rate by 15.48% when compared to the benchmarked schemes considered for investigation.

Keywords Link Aware Fault Detection · Improved Rider Optimization Algorithm · Mobile Ad hoc Network (MANET) · Link Fault Detection Rate · Black hole attack · Grey hole attack

1 Introduction

Mobile Ad hoc Network (MANET) is an independent wireless network that includes mobile nodes which enable communication without demanding a centralized Base Station (BS). A group of independent mobile nodes establish an ad hoc network without any stable infrastructure. Maintaining connectivity is indispensable in these networks as they are independent of the existing infrastructure like routers in wired networks or Access Points (APs) in infrastructure-based wireless networks. The nodes act as hosts and routers. The nodes may take part in routing by advancing packets to other nodes [1]. MANET finds its application in various areas from battlefields to user's living room. Energy constraints demand much attention as extending the lifetime of a network is challenging. The conventional schemes used for preserving energy focus on transmission energy and scheduling of active nodes in the network [2].

The main challenge faced in building a network is equipping the nodes to unceasingly preserve the information essential to appropriately route traffic. Designing effective and robust routing protocols is challenging. The nodes for forwarding data are dynamically chosen based on the algorithms used for routing and establishing network connectivity. Each node moves autonomously and subsequently changes the links with other nodes often [3, 4]. Dynamic change in the topology of the network may reduce network performance. Nevertheless, selecting multiple paths is challenging to enhance the network lifetime. To accomplish multiple routing necessities such as less delay, high Packet Delivery Ratio (PDR) and efficient adaptation to topology changes with less control overhead, methods for NP-hard optimization problem of routing are to be designed. Though the network assumes that all the nodes are cooperative and reliable, it is infeasible in real-time, as some nodes may be malevolent. Hence, it is indispensable to design a scheme that is capable of ensuring security by choosing an optimum route for data forwarding. There are chances for several types of attacks including Denial of Service (DoS), black hole attack, wormhole attack etc., Designing efficient and secure routing algorithms that facilitate finding optimal paths is an uphill task in a mobile network and has attracted the focus of many researchers.

Section 2 discusses about the works done by various researchers related to optimized selection of paths for routing in MANET. In Sect. 3, the aspects of Rider Optimization Algorithm (ROA) are discussed in detail. Section 4 shows the simulation results, while Sect. 5 gives the conclusion.

2 Related Work

In this section, the works done by various authors related to optimized routing mechanisms for MANETs are discussed.

Sheikhan and Hemmati [5] have dealt with the challenges in designing routing algorithms. Independent multi-path routing protocols focus on this issue and improve the trustworthiness, security and lifespan of network. Nevertheless, choosing an ideal multi-path is an NP-complete problem. The authors have proposed Hopfield Neural Network (HNN) wherein the parameters are enhanced using Particle Swarm Optimization (PSO) algorithm for multipath routing. Link Expiration Time (LET) between nodes is used as link consistency estimate. This scheme finds either node or link independent paths in one phase route identification. The performance of PSO-HNN routing algorithm is better in contrast to the Backup Path set Selection Algorithm (BPSA) based on path set dependability and number of paths.

Zafar and Soni [6] have designed a potential scheme for ensuring Quality of Service (QoS) using a Trust-Based QoS Protocol (TBQP) that is based on meta-heuristic Genetic Algorithm (GA) for enhancing and safeguarding MANET. GA aids in preserving QoS by selecting the fittest, that is the shortest route. Optimization schemes or meta-heuristic algorithms like GA and Neural Networks (NN) depend on Artificial Intelligence (AI), Simulated Annealing (SA) and PSO. The authors have focused on mitigating attacks against unauthorized access, data modification and impersonation attacks by malevolent nodes in the network.

Selfish nodes do not forward data packets and is usually expected to be a type of misconduct which interrupts normal network operations. Subbaraj and Savarimuthu [7] have propounded a QoS-restricted EigenTrust-based non-cooperative game model for ensuring secured and fault-tolerant perceived routing aimed at identifying secure and perceived route pairs which aids in selecting alternative paths on route failure.

Robinson and Rajaram [8] have propounded an Energy-aware Multipath routing mechanism using PSO (EMPSO) that involves incessant Recurrent Neural Network (RNN) to focus on optimization problems. Optimal paths without loops are found to deal with the link-independent paths. PSO is used for training RNN. The propounded mechanism uses dependability measures like transmission cost, energy and optimal traffic ratio amid the communicating nodes to improve the routing performance. Nodes based on enhanced link quality are involved in the route-finding phase. PSO enhances the problem by repeatedly trying to obtain an improved solution based on quality. The propounded mechanism identifies numerous loop-independent paths using PSO scheme.

Swarm Intelligence (SI)-based algorithms have demanded much attention as they can give optimized solutions guaranteeing more robustness, flexibility and reduced cost. Likewise, they are capable of dealing with large-scale refined problems without a central control entity. Ant Colony Optimization (ACO) meta-heuristic can be effectively used to balance numerous routing-associated necessities for dynamic MANETs. Zhang et al. [9] have presented a complete survey and comparison of numerous ACO-based routing schemes in MANETs. The survey includes presenting ACO principles applied in routing protocols, classifying them and relating the protocols based on design and simulation parameters and open issues.

As MANETs are infrastructure-less networks, secure data transmission is challenging in a dynamic environment. In many conventional routing mechanisms designed for MANETs, either security is not considered or only a feature of security is addressed without enhancing the routing performance. Ahmed et al. [10] have proposed Flooding Factor-based Framework for Trust Management (F3TM) for MANETs. True flooding scheme is used to classify attacker nodes based on the computation of trust value. Route discovery scheme is designed to identify an effective and safe path for data transmission using Experimental Grey Wolf Optimization (EGWO) algorithm for evaluating the nodes in the network. Enhanced Multi-Swarm Optimization (EMSO) is involved in optimizing the recognized paths.

Chintalapalli and Ananthula [11] have propounded a goal programming model called M-LionWhale based on hybrid optimization for secure routing. This optimization scheme integrates Lion Algorithm (LA) into Whale Optimization Algorithm (WOA) for optimal choice of paths in MANETs. The optimization scheme based on multiple objectives considers QoS factors such as distance, power, link lifespan delay and trust. With the assessed multi-goal features, a fitness function is established for efficient choice of paths. The performance of the propounded scheme is estimated based on metrics.

Kasthuribai and Sundararajan [12] have focused on energy-efficient multi-path routing for reliable communication by evading attackers or selfish nodes in a network. The authors have propounded reliable and QoS-dependent energy-based multi-path routing. For multi-path route choice, a PSO-based gravitational search scheme is proposed. After some transmissions, a path may include links that are not of better quality. An efficient path is selected from identified paths based on the cuckoo search algorithm. The performance parameters are compared with that of the prevailing authentic anonymous protected routing.

In a wireless network, parameters like coverage area, energy consumption, lifespan and Black Hole (BH) identification in network with interrelated nodes are considered. Shivamallaiah and Karibasappa [13] have used ACO for determining the shortest paths. ACO aids in determining additional paths when malevolent nodes exist in the network. Dual Rivest-Shamir-Adleman (DRSA) scheme and Black hole Detection (BD) are involved for recognizing malevolent nodes. For designing efficient networks, the factors are enhanced, numerous algorithms are applied to optimize the performance factors and identify the black hole in the network. The black holes are recognized based on enhanced routing algorithm in secured environment using Dual RSA. The ACO-DRSA-BD algorithm exactly identifies the BH Node (BHN) by choosing appropriate solution for data communication using ACO and secured scenario using DRSA. Hence, ACO-DRSA-BD aids in communicating data in the presence of BH attacks and improving performance.

Mohan and Reddy [14] have proposed a Whale Optimization Algorithm (WOA) used for choosing ideal trusted routes. The WOA scheme uses trust feature and distance among nodes for calculating the fitness for the route. Generally, the algorithm involves the following steps for routing which include computing trust and distance-based metrics for each node, determining k-independent paths and optimal path depending on trust and distance-based metrics. The performance of Trust-based WOA (T-Whale) is analysed based on various metrics.

Ram Mohan and Ananthula [15] have addressed security issues by propounding Jaya Cuckoo Search (JCS) optimization algorithm, which is the blend of Jaya algorithm and Cuckoo Search (CS) algorithm for introducing secure path among nodes such that the route achieved is possible and safe. The propounded algorithm is based on a fitness function which considers a multi-objective function using factors such as delay, link lifetime, trust, distance and energy in addition to the reputation factor to choose a fortified path.

Subbiah et al. [16] have focused on several issues related to security, infrastructure, severe energy sources and security. The propounded scheme emphasizes on the communication using a scheme focusing on radio station issue in route optimization which is implemented using GWO algorithm. Task-based admission control model is planned to observe the network flow and recognize the malevolent nodes. This role-based control model involves communication based on varied sized clusters. The trusts in the nodes play a dominant role and enable non-entry communication of the non-approved nodes to maintain a steady, fastened and consistent collection of nodes.

3 Rider Optimization Algorithm (ROA)

Improved Rider Optimization Algorithm-based Link Aware Fault Detection (IROA-LAFD) scheme is proposed for facilitating security in ad hoc network for mitigating grey hole and black hole attacks with enhanced link stability. This IROA-LAFD scheme targets on mitigating packet dropping in a potential manner. The steps of the proposed IROA-LAFD scheme include: (i) discovery of neighbour and route; (ii) detection of attack; (iii) analysis of link; (iv) transmission of secure packets and link fault detection using IROA Algorithm.

Initially, the network is constructed with a collection of mobile nodes in which the discovery of neighbours is achieved by broadcasting HELLO packets to the nodes at proximity. Then the mobile nodes forward Route REQuest (RREQ) packets to the mobile nodes that are within a range of 200 meters and the RREQ is broadcast to the mobile nodes that are very close to the neighbouring nodes. When the source nodes do not receive any kind of responses but receive the same RREQ packets, then the attacker nodes are identified by adding them into the list of suspects. Moreover, the path between the trustworthy nodes is established by verifying the stability of the links through the inclusion of IROA algorithm. Then, the packets are disseminated among the authenticated nodes by incorporating bogus keys. In this case, the data packets are split and transmitted based on the size of the window that forms the foundation for the construction of packets. Each packet is capable of generating the bogus key for achieving security. At this juncture, the window size is compared with the generated key. If the size of the window is greater than the key, then the hash function is included for shortening the size of the bogus key. On the other hand, the key size is increased by padding with zeros when the window size is greater than the newly created key size. Then, the key is explored with the size of the window based on the computation of outer-padded and inner-padded keys. Furthermore, the bogus keys designated as the hash keys are returned. However, the access of control and data packets by the attacker node is attained by transmitting the bogus key and data packet. The validation of node genuineness is attained through the method of pairing intermediate nodes. Furthermore, jitter is computed for identifying the link failures and attacker nodes. In addition, IROA is used for exploring the faulty links for facilitating error-free communication and improving the PDR in the network.

3.1 Discovery of Neighbour and Route

In this phase, the construction of networks is initiated and the neighbours are discovered by broadcasting HELLO packets. The HELLO packets are broadcast mainly for minimizing network traffic and overhead. In this phase, the utility of each mobile node is taken into account for neighbour and route discovery. This discovery of the route and neighbour is attained based on the network topology and configuration that exist in proximity with the source nodes that initiated communication. Then, the RREQ packets are forwarded by the source nodes to the mobile nodes that are within 200 meters of transmission range.

3.2 Detection of Attack

In this step, the attacker nodes are identified after the route is established among the mobile nodes and are stored into the list of suspects. The grey hole and black hole attackers are identified during the process of data dissemination. This proposed IROA-LAFD scheme focuses on detecting the aforementioned attacks before the commencement of data transmission. If any of these attacks are determined, then the mobile nodes that are compromised through these attacks are appended to the list of suspects. Moreover, the routes existing between the trustworthy nodes and the attacker nodes are blocked and communications are disabled.

3.3 *Analysis of Link*

Once the attacks are detected, the paths between the genuine nodes are established by using the IROA algorithm with an objective of verifying the link stability during data dissemination. It is used for choosing the optimal path based on the exploration of stability and optimization. The complete process of link stability analysis is achieved as follows.

Step 1: Initialize the nodes in the network

Step 2: Calculate relative velocity and nearest neighbour mobile nodes based on Received Signal Strength Indicator (RSSI) estimation

Step 3: Estimate the data bandwidth, response time and queue length associated with each node

Step 4: Determine the Factor of Link Quality (FLQ) based on the Eq. (1) using the data bandwidth, response time and queue length

$$\mathrm{FLQ}\,(s,d)=\sum_{i=0}^{n}\mathrm{BW}_i+\mathrm{QL}_i+\mathrm{RS}_i-\mathrm{RI}_i \tag{1}$$

Step 5: Sort the mobile nodes in the list based on the identified RSSI neighbour and FLQ parameter

Step 6: Initiate the process of link construction based on the information present in the list

Step 7: Compute the global value related to the identified RSSI neighbour and FLQ parameter for employing heuristic-based fitness search

Step 8: Update the list each time by extracting the measures of link quality

Step 9: Repeat from step 6

3.4 *Transmission of Secure Packets and Link Fault Detection Using IROA Algorithm*

The proposed scheme derives the benefits of the Improved Rider Optimization Algorithm (IROA). This IROA algorithm comprises of a collection of rider groups such as follower, bypass rider, attacker and overtaker. The riders (search agents) should follow the predefined strategy which is explained as follows.

The bypass rider has the capability of reaching the target based on the path of the lead and this rider group is initialized based on Eq. (2):

$$R_p^t = \left\{R_p^t\ (i, j)\right\}, 1 \leq i \leq N_R, 1 \leq j \leq D_{\text{RP}} \tag{2}$$

where 'N_R' and 'D_{RP}' refer to the number of riders and dimensions used in the problem of optimization at time instant 't'.

'R_p^t' pertains to the rider position any time instant 't'. In every time instant, it is inferred that the bypass riders are always in the leading positions that can be calculated based on Eq. (3):

$$R_{\text{PBye}}^{t+1} = \propto \left[R_{\text{PBye}}^t(\beta, j)^* \ \gamma(j) + R_{\text{PBye}}^t(\delta, j)^* \ [1 - \gamma(j)]\right] \tag{3}$$

where, '$\propto$' and 'β' refer to the random numbers that lie in the range of [0,1] and [0,R] respectively. Moreover, 'δ'and 'γ' are also random numbers which are varied in the range of [0. . . .R]. However, the size of the bypass rider vector (R_{PBye}^{t+1}) is of dimension '$1 \times D_{\text{RP}}$'.

The follower rider always arrives at the destination by following the bypass rider. The path traversed by the follower rider completely depends on the bypass rider. In this context, the position of the follower is updated based on Eq. (4):

$$R_{\text{PF}}^{t+1}\ (i, k) = R_{\text{PF}}^t\ (i, k) + \left\lfloor \text{Cos}\left(P_{\text{PR}} * \text{dis}_R^i * \text{ST}_{\text{Angle}}^i\right)\right\rfloor \tag{4}$$

The next rider pertaining to the category of the overtaker needs to determine its own positions as well as gather information associated with the bypass rider. But, the overtaker position depends on the factors of direction indicator, comparative success rate and selection of coordinates. Hence, the position of the overtaker search agent is updated based on Eq. (5):

$$R_{\text{PO}}^{t+1}\ (i, k) = R_{\text{PO}}^t\ (i, k) + \left\lfloor P_R^i * \text{DI}_i\right\rfloor \tag{5}$$

where 'P_R^i' and 'DI_i' are the positions of the rider and direction indicator, respectively. The value of the direction indicator is computed based on Eq. (6).

$$\text{DI}_i = \left[\frac{2}{1 - \log\left(R_{\text{SR}}^i\right)}\right] - 1 \tag{6}$$

The last category of rider named attacker has the potential of reaching the objective point, and it determines the leader position and strategy of travelling, which is very similar to the phenomenal characteristics of follower. In particular, the positions of the attackers are updated based on Eq. (7):

$$R_{\text{PA}}^{t+1}\ (i, k) = R_{\text{PA}}^t\ (i, k) + \left\lfloor \text{Cos}\ \left(P_{\text{PR}} * \text{dis}_R^i * \text{ST}_{\text{Angle}}^i\right)\right\rfloor \tag{7}$$

However, the basic ROA scheme has some limitations in delayed convergence and stagnation during the exploitation and exploration phases. Thus, ROA is improved using the strategy of gravitational search. In this improved ROA, the association between the different categories of rider search agents is considered for link fault detection. To represent the role of gravitational search in ROA, the rider position can be defined by Eq. (8):

$$R_P^i = \left(R_{P_1}^1, R_{P_i}^2, \ldots\ldots R_{P_i}^{D_N}, \ldots\ldots R_{P_i}^{N_R}\right) \tag{8}$$

In this situation, the acting force is defined in Eq. (9):

$$F_{\mathrm{Aij}}^t = S_{\mathrm{Const}} \frac{m_i * m_j}{d_{\mathrm{rider}(i,j)}} \left(R_{\mathrm{Pj}}^t - R_{\mathrm{Pi}}^t\right) \tag{9}$$

Based on the aforementioned equation, the positions associated with the bypass rider, follower overtaker and attacker are updated based on Eqs. (10–13):

$$R_{\mathrm{PBye}}^{t+1} = \propto \left[R_{P_{\mathrm{Bye}}}^t(\beta, j)^* \, \gamma(j) + R_{P_{\mathrm{Bye}}}^{\mathrm{t}}(\delta, j)^* \, [1 - \gamma(j)]\right] + \mathrm{acc}_{\mathrm{Bye}} \tag{10}$$

$$R_{\mathrm{PF}}^{t+1}(i,k) = R_{\mathrm{PF}}^t(i,k) + \left\lfloor \mathrm{Cos}\left(P_{\mathrm{PR}} * \mathrm{dis}_R^i * \mathrm{ST}_{\mathrm{Angle}}^i\right)\right\rfloor + \mathrm{acc}_{\mathrm{Follower}} \tag{11}$$

$$R_{\mathrm{PO}}^{t+1}(i,k) = R_{\mathrm{PO}}^t(\mathrm{i,k}) + \left\lfloor P_R^i * \mathrm{DI}_i \right\rfloor + \mathrm{acc}_{\mathrm{Overtaker}} \tag{12}$$

$$R_{\mathrm{PA}}^{t+1}(i,k) = R_{\mathrm{PA}}^t(i,k) + \left\lfloor \mathrm{Cos}\left(P_{\mathrm{PR}} * \mathrm{dis}_R^i * \mathrm{ST}_{\mathrm{Angle}}^i\right)\right\rfloor + \mathrm{acc}_{\mathrm{Attacker}} \tag{13}$$

where, '$\mathrm{acc}_{\mathrm{Bye}}$', '$\mathrm{acc}_{\mathrm{Follower}}$', '$\mathrm{acc}_{\mathrm{Overtaker}}$' and '$\mathrm{acc}_{\mathrm{Attacker}}$' are the accelerations introduced into the process of improved rider optimization algorithm.

4 Simulation Results and Discussion

The propounded Improved Rider Optimization Algorithm-based Link Aware Fault Detection (IROA-LAFD) and the benchmarked Energy-aware Multipath Routing mechanism using PSO (EMPSO), M-LionWhale and Framework for Trust Management (F3TM) schemes are simulated using ns-2.35. The simulation parameters taken for implementation of the propounded scheme are shown in Table 1. The

Table 1 Simulation parameters

Parameters	Values
Simulation time	250 s
Topology size	1000 × 1200 m^2
Number of nodes	500
Maximum power	1 W
Pause time	5–8 s
Maximum speed	10 m/s
Traffic type	CBR/VBR
Packet size	1024 bytes
Wireless channel capacity	2 Mbps
Routing protocol	AODV and OLSR
Transmission range	250 m (limited 200 m)
Mobility model	Random waypoint
Wireless standard	802.11b

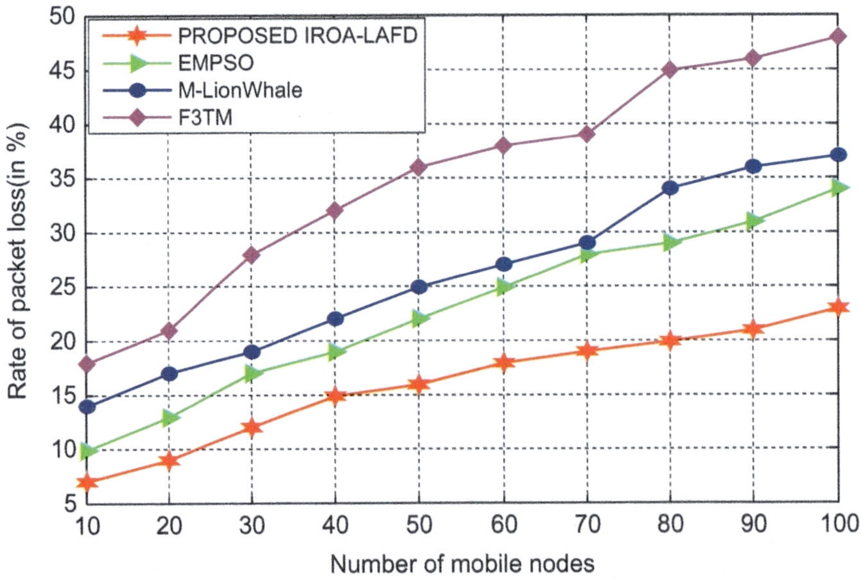

Fig. 1 Rate of packet loss of the proposed IROA-LAFD scheme for varying number of nodes

performance of the proposed IROA-LAFD is assessed in terms of rate of packet loss, rate of link failure detection, end-to-end delay, latency, PDR and throughput for varying number of mobile nodes.

In Fig. 1, it is seen that the proposed IROA-LAFD scheme involves 12.6%, 14.2% and 16.2% less rate of packet loss in contrast to the benchmarked EMPSO, M-LionWhale and F3TM schemes considered for comparison. From Fig. 2, it is evident that the propounded IROA-LAFD offers 9.7%, 12.2% and 15.3% improved link

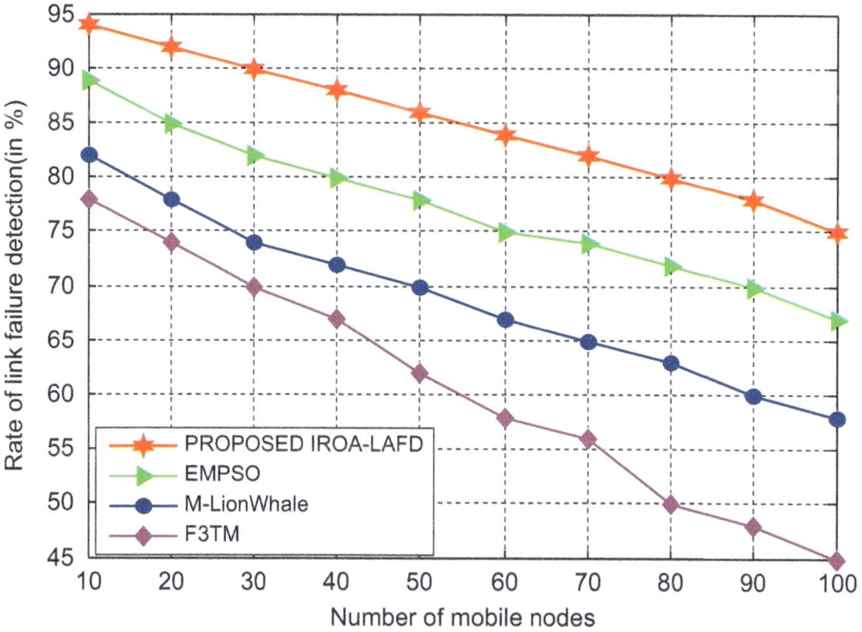

Fig. 2 Rate of link failure detection of the proposed IROA-LAFD scheme for varying number of nodes

failure detection when compared to the benchmarked EMPSO, M-LionWhale and F3TM schemes.

From Fig. 3, it is evident that the proposed IROA-LAFD involves 10.2%, 13.9% and 15.6% less end-to-end delay in contrast to the benchmarked EMPSO, M-LionWhale and F3TM schemes considered for comparison.

In Fig. 4, it is seen that the benchmarked EMPSO, M-LionWhale and F3TM schemes involve 11.9%, 14.5% and 17.1% more latency in contrast to the proposed IROA-LAFD scheme. From Fig. 5, it is obvious that the benchmarked EMPSO, M-LionWhale and F3TM schemes offer 12.3%, 15.7% and 16.4% less PDR when compared to the proposed IROA-LAFD scheme.

Finally, from Fig. 6, it is clear that the proposed IROA-LAFD scheme offers 10.5%, 13.2% and 15.7% better throughput when compared to the benchmarked EMPSO, M-LionWhale and F3TM schemes.

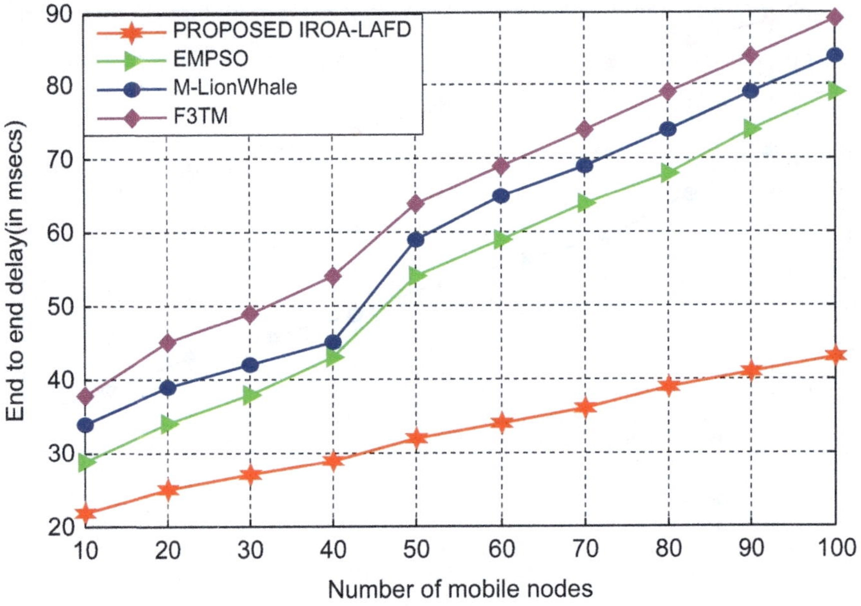

Fig. 3 End-to-end delay of the proposed IROA-LAFD scheme for varying number of nodes

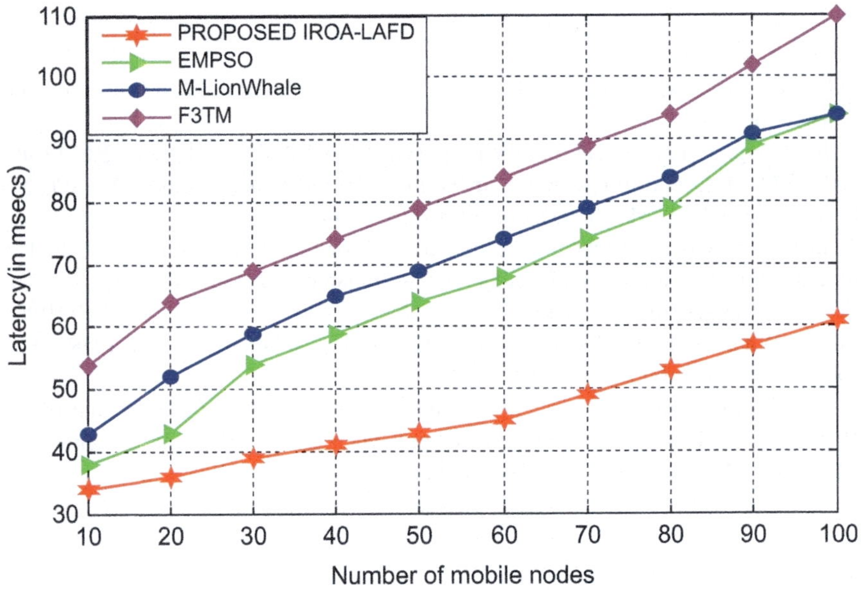

Fig. 4 Latency of the proposed IROA-LAFD scheme for varying number of nodes

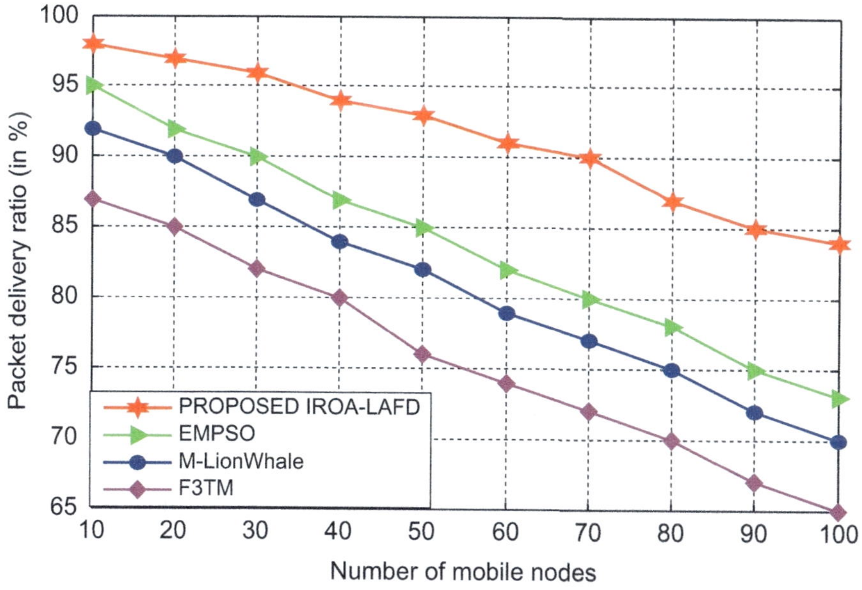

Fig. 5 Packet delivery ratio of the proposed IROA-LAFD scheme for varying number of nodes

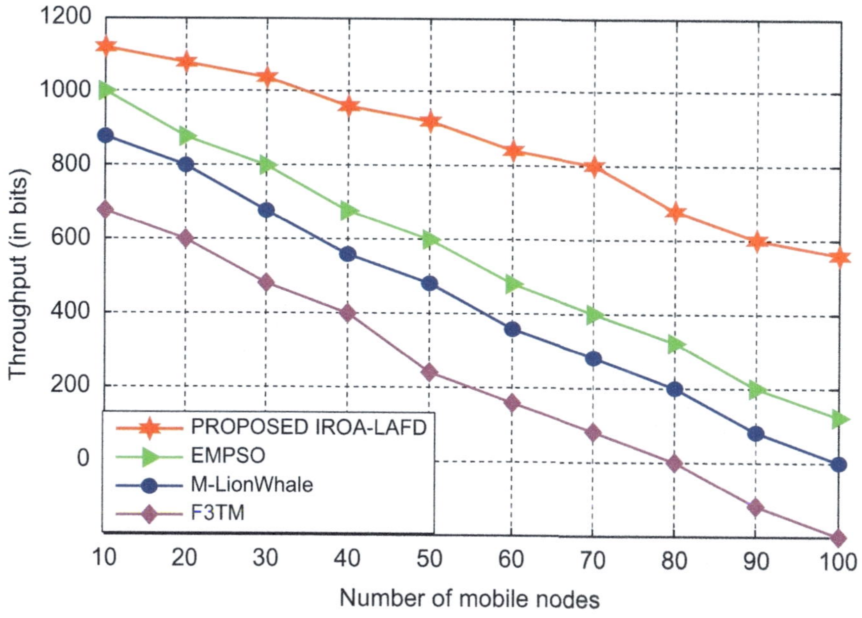

Fig. 6 Throughput of the proposed IROA-LAFD scheme for varying number of nodes

5 Conclusion

This proposed Improved Rider Optimization Algorithm-based Link Aware Fault Detection (IROA-LAFD) scheme is presented as a reliable attempt for ensuring security against the grey hole and black hole attacks by exploring possible dimensions of link stability. This proposed IROA-LAFD scheme calculates the relative velocity and nearest neighbour mobile nodes based on RSSI estimation in addition to data bandwidth, response time and queue length associated with each node with an objective of estimating the Factor of Link Quality (FLQ). FLQ aids in better exploration of links in order to identify whether they are being influenced by the grey hole and black hole attacks during the data dissemination process. This IROA-LAFD scheme includes the gravitational search ROA algorithm for exploring the link paths established between the trust nodes of the network. The simulation experiments of the proposed IROA-LAFD scheme is identified to minimize the packet loss by 14.28%, end-to-end delay by 12.94% and latency by 13.96% when compared to the benchmarked schemes considered for investigation. In the near future, it is also planned to formulate a Firefly platonic search-based link path quality estimation for comparing its performance with the proposed scheme.

References

1. Deva, P. M., & Priyanka, P. (2015). PPCLSS: Probabilistic prediction coefficient link stability scheme based routing in MANETs. *International Journal of Computer Science & Engineering Technology, 6*(4), 246–256.
2. Deva, P. M., Aishwarya, R., Anushya, S., & Keerthana, S. (2017). Detection and avoidance of malicious nodes in MANETs. *International Journal of Scientific Research in Computer Science, Engineering and Information Technology, 2*(2), 434–441.
3. Deva, P. M., Janakiraman, S., Sandhya, G., & Aishwaryalakshmi, G. (2019). Efficient pre-authentication scheme for inter-ASN handover in high mobility MANET. In *Wireless networks* (pp. 1–15). Springer. https://doi.org/10.1007/s11276-019-02185-2.
4. Maharaja, S., Jeyalakshmi, R., Sabarish Kanna, A. V., & Deva, P. M. (2019). Secured routing in mobile ad hoc networks (MANETs). *International Journal of Scientific Research in Computer Science, Engineering and Information Technology, 5*(2), 277–289.
5. Sheikhan, M., & Hemmati, E. (2012). PSO-optimized Hopfield neural network-based multipath routing for mobile ad-hoc networks. *International Journal of Computational Intelligence Systems, 5*(3), 568–581.
6. Zafar, S., & Soni, M. K. (2014). Trust based QOS protocol (TBQP) using meta-heuristic genetic algorithm for optimizing and securing MANET. In *IEEE international conference on reliability optimization and information technology* (pp. 173–177).
7. Subbaraj, S., & Savarimuthu, P. (2014). EigenTrust-based non-cooperative game model assisting ACO look-ahead secure routing against selfishness. *EURASIP Journal on Wireless Communications and Networking, 2014*(1), 78.
8. Robinson, Y. H., & Rajaram, M. (2015). Energy-aware multipath routing scheme based on particle swarm optimization in mobile ad hoc networks. *The Scientific World Journal, 2015*, 284276.

9. Zhang, H., Wang, X., Memarmoshrefi, P., & Hogrefe, D. (2017). A survey of ant colony optimization based routing protocols for mobile ad hoc networks. *IEEE Access, 5*, 24139–24161.
10. Ahmed, M. N., Abdullah, A. H., Chizari, H., & Kaiwartya, O. (2017). F3TM: Flooding factor based trust management framework for secure data transmission in MANETs. *Journal of King Saud University-Computer and Information Sciences, 29*(3), 269–280.
11. Chintalapalli, R. M., & Ananthula, V. R. (2018). M-LionWhale: Multi-objective optimisation model for secure routing in mobile ad-hoc network. *IET Communications, 12*(12), 1406–1415.
12. Kasthuribai, P. T., & Sundararajan, M. (2018). Secured and QoS based energy-aware multipath routing in MANET. *Wireless Personal Communications, 101*(4), 2349–2364.
13. Shivamallaiah, S. M., & Karibasappa, K. (2018). An efficient detection of BH attack with secured routing using ACO and DualRSA in MANETs. *International Journal of Intelligent Engineering and Systems, 11*(2), 246–255.
14. Mohan, C. R., & Reddy, A. V. (2018). T-whale: Trust and whale optimization model for secure routing in mobile ad-hoc network. *International Journal of Artificial Life Research, 8*(2), 67–79.
15. Ram Mohan, C., & Ananthula, V. R. (2019). Reputation-based secure routing protocol in mobile ad-hoc network using Jaya Cuckoo optimization. *International Journal of Modeling, Simulation, and Scientific Computing, 10*(03), 1950014.
16. Subbiah, B., Obaidat, M. S., Sriram, S., Manoharn, R., & Chandrasekaran, S. K. (2020). Selection of intermediate routes for secure data communication systems using graph theory application and grey wolf optimisation algorithm in MANETs. *IET Networks*. https://doi.org/10.1049/iet-net.2020.0051.

Media Access Protocol for Wireless Sensor Network Using Active Reception Scheme-Based Energy-Efficient Technique

Anushree Goud and Bindu Garg

Abstract Reception Media Access Control is a TDMA-based MAC protocol for data transmission in WSNs. The feature is to preserve energy through restrain traffic overhearing, in-active listening, retransmission of data and packet collision. R-MAC uses a distributed scheduling procedure to assign a time slot for reception called Reception Slot (RS) to each sensor, and share message of reception slot to each of its neighbouring device. A sensor who has data to transmit can consequently wake up in reception slot of its intended receiver only. R-MAC guarantees that only one one-hop neighbour can transmit at a time in reception slot of its intended receiver, which assures collision avoidance. The result analysis of R-MAC is analysed with respect to power consumption, traffic and message delay in multi-hop networks through detailed simulation. The result of R-MAC is compared against SMAC. Network simulator 2 (NS-2) is used for simulation.

Keywords Reception Slot (RS) · Δ-slot · Slot request · Slot response · Data request · Data response

1 Introduction

WSNs consist of small individual sensors coordinating among themselves to analyse environmental conditions, such as pressure, sound, temperature and vibrations [1]. Sensors in WSNs have sensing, computation and wireless transfer capabilities. WSNs have emerged as a substitute of large network infrastructures for a large number of application, for example, precision agriculture, habitat monitoring,

A. Goud (✉)
Narsee Monjee Institute of Management studies, Mumbai, Maharashtra, India
e-mail: anushree.goud@nmims.edu

B. Garg
Bharati Vidyapeeth College of Engineering, Pune, India
e-mail: brgarg@bvucoep.edu.in

A. Haldorai et al. (eds.), *3rd EAI International Conference on Big Data Innovation for Sustainable Cognitive Computing*, EAI/Springer Innovations in Communication and Computing, https://doi.org/10.1007/978-3-030-78750-9_12

structure monitoring, environmental and medical diagnostics. Detecting complex scenario interactions is the most dramatic applications of WSNs.

Designing MAC scheme for WSNs involves many challenges. Energy consumption is one of the most vital issue in WSN as sensors have limited battery power, and it is very hard or sometimes impossible to recharge or replace batteries of sensors. Our main aim is to consider the sources of energy waste while designing an efficient protocol. Idle listening is the first cause of unwanted energy wastage. It occurs usually when the node is listening to the environment for expected traffic but there is no traffic in the channel.

Collision is the second source of unwanted energy wastage. When multiple messages are transmitted in the same channel simultaneously, collision may occur and they will be of no use and must be discarded. Retransmission of collided messages consumes energy. Collision is a vital issue to consider in contention-based MAC scheme. Next source is overhearing, which occurs when a sensor listens to the traffic that is destined to separate sensors. Where traffic load is high, overhearing can be a serious issue to deployed network. Sending and receiving control packets and control packet overhearing are other sources of energy waste as it does not directly convey useful data.

In this research work, MAC scheme is proposed for data transmission in WSNs, which is energy efficient. This solution is suited for network scenarios where sensor mobility is low, traffic rate is low-to-moderate. Environmental monitoring, structure monitoring and precision agriculture are some examples of such applications. Our proposed MAC scheme limits the energy waste through limiting overhearing, packet collision, retransmission, idle listening and packet over-emitting. R-MAC is a time division multiple access-based protocol, which uses a distributed procedure that allocates duration to each node for data and spreads the information about allocated time slot among neighbouring sensors.

The chapter is structured as follows: In Sect. 2, literature review is presented, Sect. 3 describes the proposed MAC scheme referred to as R-MAC. Section 4 contains the simulation results followed by conclusions in further section.

2 Literature Review

In literature, many MAC protocols have been proposed, which can be categorised as contention-free and contention-based scheme. It offers low traffic rate scenarios, since they mainly work on the basis of Carrier Sense Multiple Access (CSMA) or Carrier Sense Multiple Access/Collision Avoidance (CSMA/CA), allowing collisions to occur, and try to recover from these collisions by using back-off algorithm. Examples of contention-based protocols are S-MAC (Sensor MAC) [6, 13], D-MAC (Data-gathering MAC) [7, 16], T-MAC (Timeout MAC) [8, 15]. S-MAC implements periodical listening and sleeping mechanism to minimize energy waste. In other word, each node goes to sleep for some time, and then wakes up and listens to see if any other node wants to transmit to it [6].

On the other side, contention-free scheduled MAC protocols use TDMA slotted structure, where time slots are allocated to the nodes in the network. Energy efficiency is achieved by ensuring that each node is asleep whenever it is not scheduled to transmit or receive. Examples of contention-free MAC schemes are SPARE-MAC [9], TDMA-ASAP [10], TRAMA [11], DEE-MAC [12]. A reception schedule is a time slot during which a sensor becomes active to receive data. This conserves energy by limiting overhearing and idle listening at the expense of increased collision probability [9]. Collision happens when multiple nodes transmit to the same node in the same reception schedule [14]. does not use collision avoidance, but it recovers after collisions happen.

3 R-MAC Description

R-MAC is a TDMA-based scheme, where individual sensor is having a time slot for reception called Reception Slot (RS), and sensors know the RS of all its receivers. A RS is a time slot when sensor node becomes active for exchange of data. The rationale behind R-MAC is that if a sensor needs to transmit, the receiving sensor restrains its all other one-hop neighbours from transmitting during ongoing transmission to avoid collision. Furthermore, time synchronization among sensors is the primary requirement of R-MAC, which can be achieved through different proposed Schemes [2–5].

In the next section, frame structure of R-MAC is presented. The reception slot assignment procedure is followed by data transmission.

3.1 Frame Structure

R-MAC describes the frame structure shown in Figs. 1 and 2. Each is divided into N Reception Slots (RSs). Each RS is further divided into N + 3 Δ-slots and a data slot. Δ-slots are used for coordination among neighbouring sensors. On the other hand, data slot is reserved for data reception. The parameter N depends on the traffic requirements and topology of network. Δ-slots forming 1 to N are assigned to the neighbouring sensors. Following are Δ-slot assignment requirements:

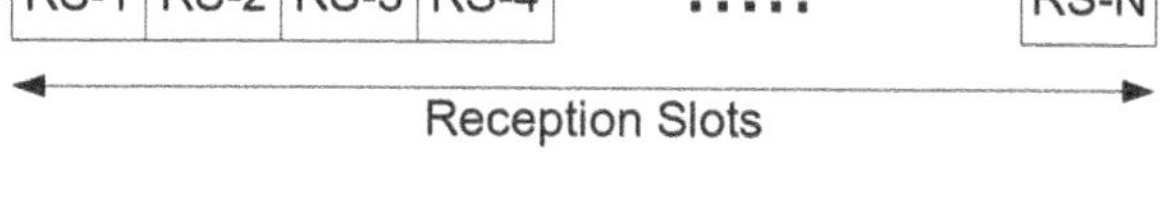

Fig. 1 Frame structure

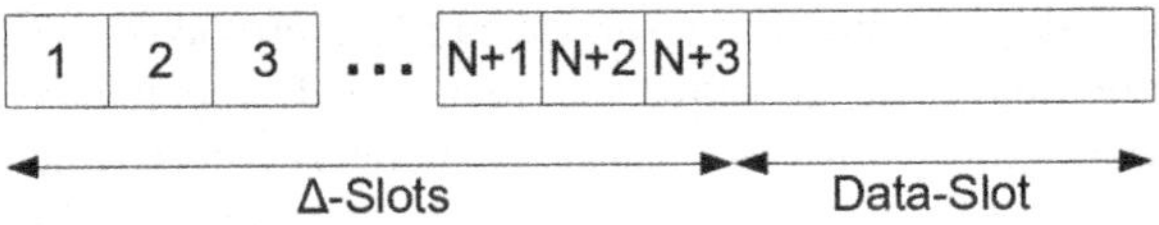

Fig. 2 Reception slot structure

- Each sensor is assigned a Δ-slot in all its potential receivers' RS.
- Each sensor knows its Δ-slot in all its potential receivers' RS.

3.2 Reception Slot Assignment

Next step after activation of a sensor is to gain a reception slot. Δ-slots N + 2 and N + 3 in each RS are reserved from reception slot assignment. Reception slot assignment completes in three steps.

4 Step 1

After activation, the newcomer sensor sends a slot request in N + 2th Δ-slot of every RS. A slot request contains sender's RS number and Δ-slot number of receiver in sender's RS. In step 1, slot request contains NULL for both sender's RS number and Δ-slot number of receiver *in sender's RS.*

Rule 1: *Each sensor remains active during all Δ-slots of its own RS.*

Each receiver of slot request checks Δ-slot number of sender in its own RS, if Δ-slot is not assigned yet, assigns a Δ-slot in its own RS to the sender and sends a slot-response in N + 3th Δ-slot. A slot response contains sender's RS number and receiver's Δ-slot number in sender's RS. If newcomer receives no slot response in N + 3th Δ-slot, marks the current RS as AVAILABLE. If newcomer receives one slot response, marks the current RS RESERVER, stores the RS number of the sender and its own Δ-slot number in sender's RS and assigns a Δ-slot to the sender in its own RS. Collision may happen at newcomer where multiple nodes send slot response on the same RS. If collision happens at newcomer, it marks the corresponding RS COLLIDE.

Rule 2: *An AVAILABLE reception slot must not interfere with any other reception slot of one-hop neighbours.*
Rule 3: *Two one-hop neighbours cannot have same reception slot.*

After step 1, newcomer acquires a RS from the list of *AVAILABLE* reception slots.

5 Step2

In the following frame, newcomer sends *slot requests* in $N + 2th$ Δ-slot of each *RESERVED* and *COLLIDE* reception slot. If RS is coded *COLLIDE,* the value of *Δ-slot number of receiver in sender's RS* in *slot request* is set *NULL.* At the receiver end, receiver stores the sender's RS number. If the value of Δ-*slot number of receiver in sender's RS* is not *NULL, it* stores in its own Δ-slot number in sender's RS and sends a slot response in $N + 3th$ Δ-slot.

The NULL value of Δ-*slot number of receiver in sender's RS* is interpreted as the indication of collision.

6 Step 3

Each collision-occurring node refrains from exchanging *slot request* for a number of frames, interpreted as *i*

$$i = \begin{cases} \text{random}\left[1, 2^k\right] & \text{, If } k \leqslant 10 \\ \text{random}\left[1, 2^{10}\right] & \text{, otherwise} \end{cases}$$

where k is the number of consecutive collisions.

After *i* frames, the receiver sends slot request in $N + 2th$ Δ-slot of newcomer's RS. After receiving *slot request,* newcomer stores RS number of sender and its own Δ-slot number in sender's RS, assigns a Δ-slot to the sender in its own RS and sends a *slot-response.* Receiver of slot response stores its own Δ-slot number in sender's RS.

Figure 3 shows the case of a new sensor entering the system (sensor 3). Sensor 3 waits for the next frame and sends out a *slot request* in sixth Δ-slot of RS-1 which is acquired by sensor 1. Sensor 1 assigns a Δ-slot to sensor 3 in its own RS (RS-1) and sends out *slot response* in seventh Δ-slot. Sensor 3 receives *slot response* from sensor 1 containing RS of sensor 1 and Δ-slot of sensor 3 in RS of sensor 1. Sensor 3 assign a Δ-slot to sensor 1 in its own RS and marks RS-1 *RESERVED.* Same procedure is followed in RS-2 which is owned by sensor 2. Sensor 3 sends *slot request* in sixth Δ-slot of RS-3 and RS-4 but does not get *slot response* in seventh Δ-slot of RS-3 and RS-4. So, sensor 3 marks RS-3 and RS-4 *AVAILABLE.* After frame 1, sensor 3 acquires a RS from the list of *AVAILABLE* reception slots. In the next frame (frame 2), sensor 3 sends out a *slot request* in sixth Δ-slot of RS-1 to sensor 1 containing RS number of sensor 3 and Δ-slot number of sensor 1 in the RS of sensor 3. Sensor 1 sends a *slot response* in seventh Δ-slot of RS-1 to sensor 3. Same procedure is followed in RS-2. Sensor 3 does not send slot requests in RS-3 and RS-4 because they are coded *AVAILABLE.* After frame 2, sensor 3 has acquired a RS and Δ-slot in each reception slot of its neighbours and each neighbour of sensor 3 knows RS number of sensor 3 and has a Δ-slot in RS of sensor 3.

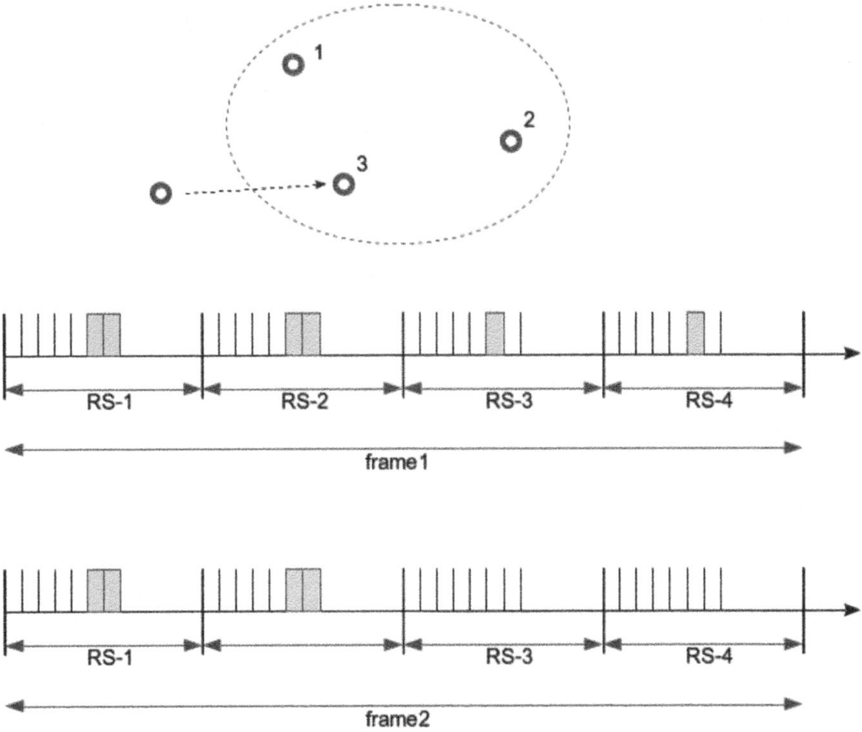

Fig. 3 Reception slot assignment procedure

6.1 Data Transfer

Once a sensor has gained a reception slot mentioned in the procedure explained, it can receive and transmit data. For this purpose, it contains information of one-hop neighbours with the corresponding reception slot and its own Δ-slot in its one-hop neighbour's RS.

Rule 4: *A node can transfer data request only in its own assigned Δ-slot in receiver's RS.*

At the receiver end, every node wakes up in its own reception slot and receives data request. As soon as data request is received by its intended receiver, the receiver switches off its radio and sets timer to become active in N + 1th Δ-slot of current RS. All data requests except first data request intended to the same destination are not received as the radio of the receiver is switched off. In N + 1th Δ-slot of current RS, receiver sends data response which announces the winner of the following data slot of current RS. In other words, data response is destined to the winner of the following data slot of current RS but it is also received by all one-hop neighbours who want to transmit in current RS. The winner of data slot starts sending in following data slot of current RS and all other one-hop neighbours of receiver refrain from transmitting in current RS.

R-MAC does not allow two nodes to transmit to a node in one RS. If a sensor does not get any data request in first N Δ-slots, switch off its radio for the entire data slot.

The idea behind R-MAC is to conserve energy through keeping the node's radio off while it is neither receiving nor transmitting. To achieve this, a sensor turns on its radio only:

- During its Δ-slot, $N + 1th$ Δ-slot and data slot in the reception slot of the receiver when it is ready to transmit.
- During its reception slot.
- **Data-request collision at receiver:** Collision can occur at the receiver if it overhears a data request destined to any other sensor and receives a data request destined to it. After collision, receiver waits for data request in the following Δ-slots. If it receives any data request in following Δ-slots than send a data response in N + 1th Δ-slot, otherwise switch off its radio for the entire data slot.

For illustration, consider the network shown in Fig. 4, where sensor 1 and 2 have same RS (according to **Rule 3,** sensor 1 and 2 can have same RS as they are two-hop neighbours). Δ-slot number of sensor 3 in RS of sensor 1 is same as Δ-slot number of sensor 4 in RS of sensor 2.Assume that sensor 3 has traffic to transmit to sensor 1, and sensor 4 has traffic to transmit to sensor 2. Because sensor 3 is in the range of sensor 2, if sensor 3 and 4 were to transmit in the same frame (node 1 and 2 has same RS) there would be collision at sensor 2. Hence, only sensor 1 transmits data response and sensor 3 subsequently transmits in the following data slot whereas sensor 4 will have to wait for next frame.

Data-Response Collision at Sender Collision can occur at the sender if it overhears a data response destined to any other sensor and receives a data response destined to it. After collision, the sender refrains from transmitting in current RS and retries in the next frame.

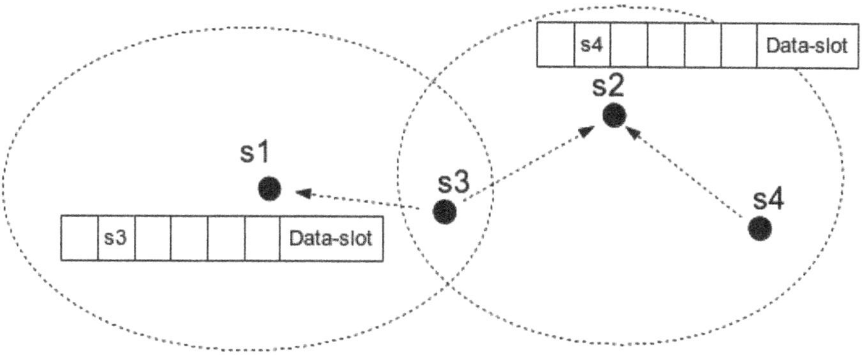

Fig. 4 Data-request collision at receiver

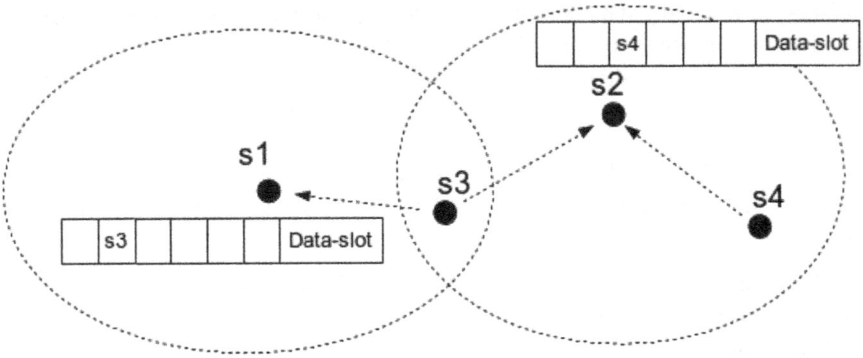

Fig. 5 Data-response collision at sender

For illustration, consider the network shown in Fig. 5, where sensors 1 and 2 have same RS. Δ-slot number of sensor 3 in RS of sensor 1 is different from Δ-slot number of sensor 4 in RS of sensor 2. Assume that sensor 3 has traffic to transmit to sensor 1, and sensor 4 has traffic to transmit to sensor 2. If both sensor 3 and 4 try to transmit in same frame, both sensor 3 and 4 will send data request in their corresponding Δ-slots. Sensor 1 will send data response in N + 1th Δ-slot to sensor 3, and sensor 2 will send slot response in the same Δ-slot to sensor 4 which will be overheard by sensor 3. Hence, there will be collision at sensor 3, and it will refrain from transmitting in current RS and sensor 4 can transmit to sensor 2 in current RS.

7 Evaluation

7.1 Simulation Environment

To test the performance of R-MAC, simulation analysis is conducted in two different scenarios as shown in Fig. 6 [9]. The first one, Fig. 6a is a fully connected cluster where every individual node is in the range of every other node in the network. The second topology (Fig. 6b) is a tree topology where traffic passes from leaf to a data sink node.

The analysis of Reception-based Media Access Control is assessed using three performance metrics: offered traffic, consumed power and end-to-end delivery delay. Table 1 contains the standard simulation parameters.

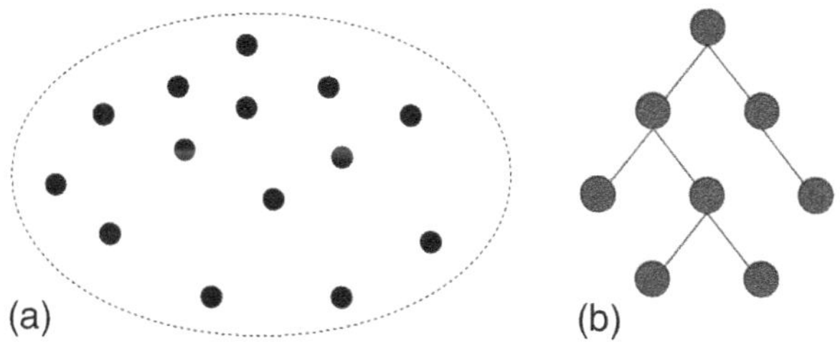

Fig. 6 (**a**) Fully connected cluster topology, (**b**) tree topology

Table 1 Simulation parameter setting

Parameter	Value
Simulation time	10,000 s
Bandwidth	250Kbps
Data slot	565byte
Δ-slot	64byte
Packet length	512byte
Transmission power	24 mW
Reception power	13.5 mW
Idle power	13.5 mW
Sleep power	5 μW

7.2 *Fully Connected Cluster Topology*

Figure 7 shows the comparison between average delivery delay measured for different values of *N*. Simulation results shows that average delivery delay increases very slowly as offered traffic is increased. Figure 6 also shows that average delivery delay is directly proportional to the value of *N*.

Figure 8 shows the comparison between average power consumption for different values of N in fully connected cluster topology. Average power consumption is inversely proportional to the value of *N*.

Performance of R-MAC is also compared with SPARE-MAC, which is a TDMA-based protocol. Bandwidth and packet length are the same as given in Table 1 for simulation of SPARE-MAC (Fig. 9).

In Fig. 10, comparison of SPARE-MAC and R-MAC is shown for average consumed power versus achieved throughput.

A. *Tree Topology.*

In this section, simulation results are presented for networks where sensors are organized in two-level tree topology, leaf sensors (sensors at level 2) transmit to data sink (sensors at level 0) (Figs. 11 and 12).

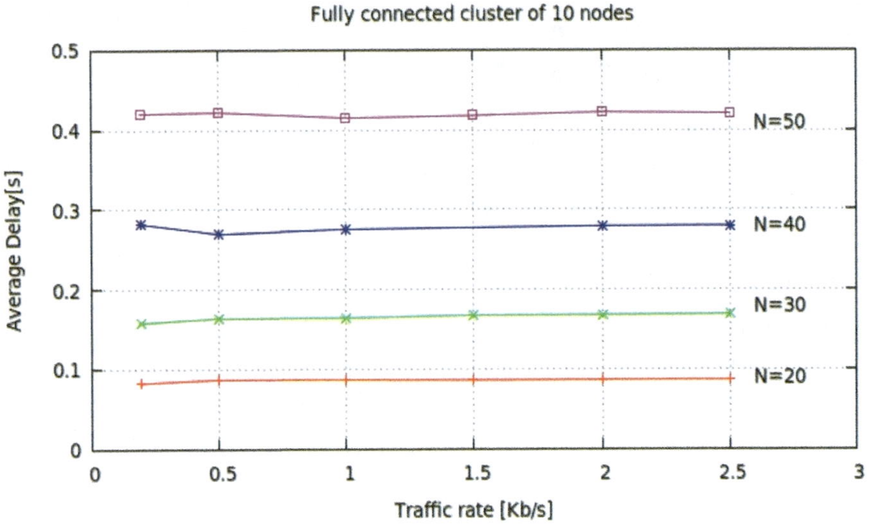

Fig. 7 Average delivery delay versus the offered traffic rate

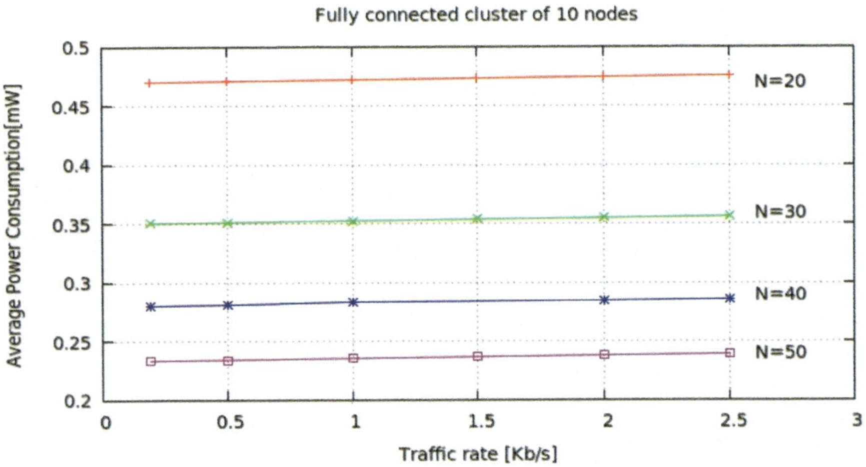

Fig. 8 Average power consumption versus the offered traffic

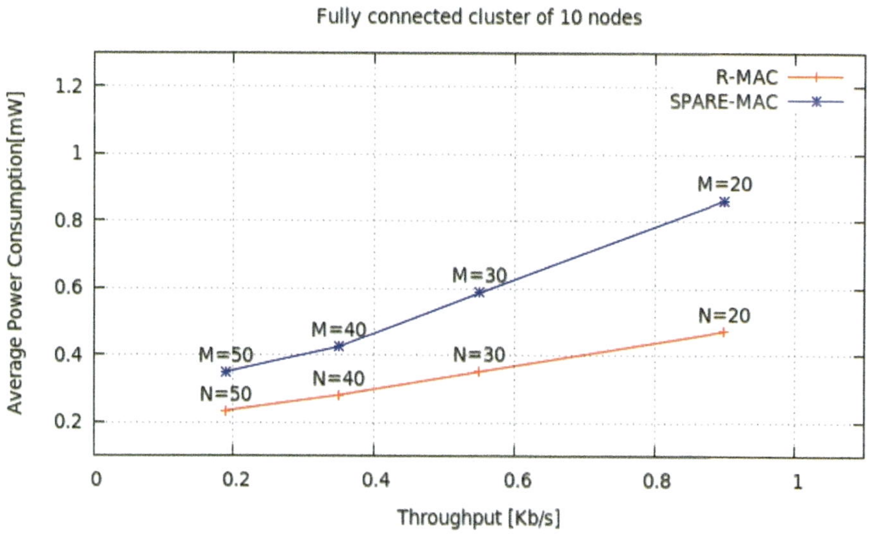

Fig. 9 Comparison between SPARE-MAC and R-MAC. Average delay versus achieved throughput in cluster topology [14]

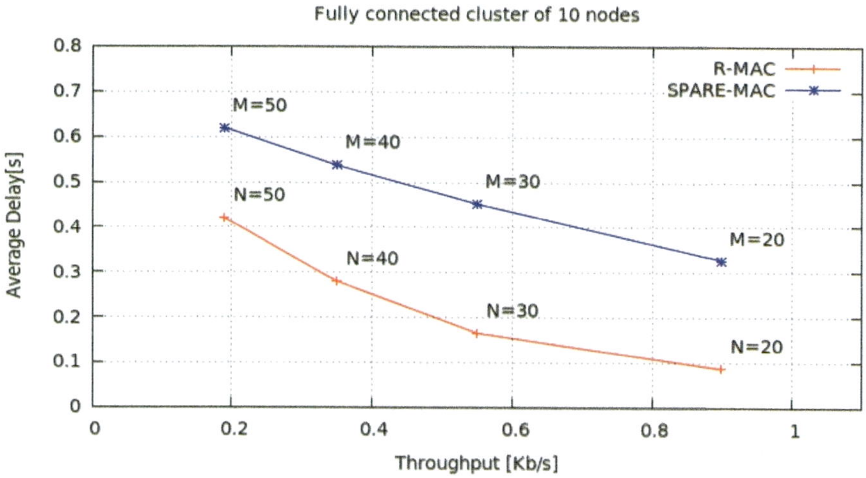

Fig. 10 Comparison between SPARE-MAC and R-MAC. Average consumed power versus achieved throughput

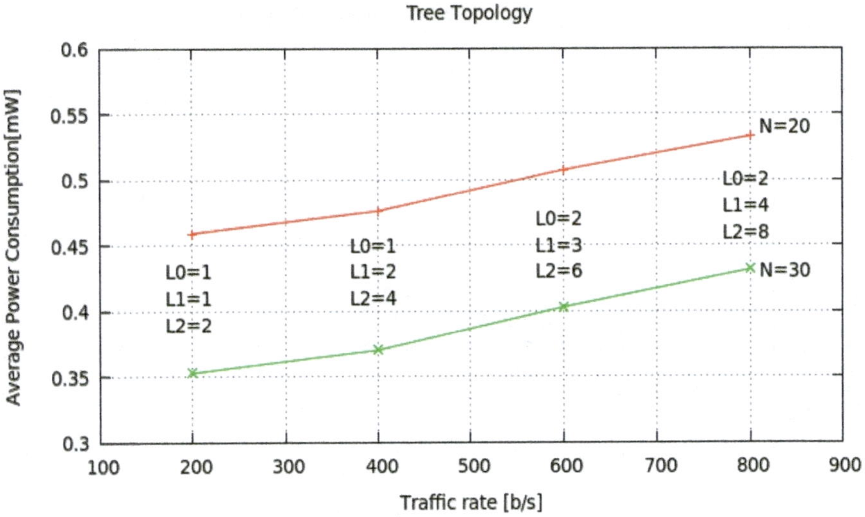

Fig. 11 Average power consumption versus traffic rate at data sink in tree topology (Figure 6b is a tree topology. It is average power consumption versus traffic rate)

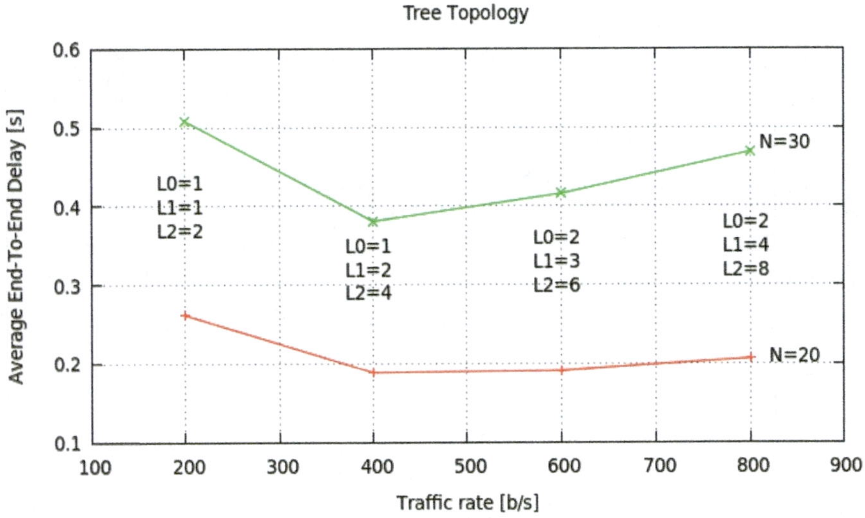

Fig. 12 Average delay versus traffic rate in bits per second at data sink in tree topology (Figure 6b is a tree topology. It is average delay vs traffic rate)

8 Conclusion

The work R-MAC is an energy-efficient Media Access Control approach, which is proposed for message transmission in Wireless Sensor Network and shown improvement in different energy levels.

References

1. Akyildiz, I. F., Su, W., Sankarasubramaniam, Y., & Cayirci, E. (2002). A survey of sensor networks. *Proceedings of IEEE Communication Magazine.*
2. Rowe, A., Mangharam, R., & Rajkumar, R. (2006). RT-link: A time-synchronized link protocol for energy constrained multi-hop wireless networks. *Proceedings of IEEE SECON*, 402–411.
3. Mangharam, R., Rowe, A., & Rajkumar, R. (2006). FireFly: A time synchronized scalable real-time sensor network Plateform. *Proceedings Of IEEE SECON.*
4. Elson, J., & Estrin, D. (2001). Time synchronization for wireless sensor networks. *Proceedings of IEEE IPDPS*, 1965–1970.
5. Ganeriwal, S., Kumar, R., & Srivastava, M. (2003). Timing-sync protocol for sensor networks. *Proceedings Of ACM SenSys*, 138–149.
6. Ye, W., Heidemann, J., & Estrin, D. (2002). An energy-efficient MAC protocol for wireless sensor networks. *Proceedings of IEEE INFOCOM.*
7. Lu, G., Krishnamachari, B., & Raghavendra, C. (2004). An adaptive energy-efficient and low-latency MAC for data-gathering in sensor networks. In *Proceedings of the 4th international IEEE workshop on algorithms for wireless, Mobile, ad hoc and sensor networks (WMAN).*
8. Dam, T. V., & Langendoen, K. (2003). An adaptive energy-efficient MAC protocol for wireless sensor networks. *Proceedings of ACM SenSys.*
9. Campelli, L., Capone, A., Cesana, M., & Ekici, E. (2007). *A receiver oriented MAC protocol for wireless sensor networks.* IEEE.
10. Gobriel, S., Mosse, D., & Cleric, R. (2009). *Sensor network TDMA scheduling with adaptive slot-stealing and parallelism.* IEEE.
11. Rajendran, V., Obraczka, K., & Garcia-Luna-Aceves, J. J. (2003). Energy-efficient, collision-free medium access control for wireless sensor networks. *Proceedings of ACM Sensys.*
12. Cho, S., Kanuri, K., Cho, J.-W., Lee, J.-Y., & June, S.-D. (2005). Dynamic energy efficient TDMA-based MAC protocol for wireless sensor networks. *Proceedings of ICAS/ICNS/IEEE.*
13. Ye, W., Heidemann, J., & Estrin, D. (2004). Medium access control with coordinated, adaptive sleeping for wireless sensor networks. *ACM/IEEE Transactions on Networking, 12*(3), 493–506.
14. Turati, F., Cesana, M., & Campelli, L. (2009). SPARE MAC enhanced: A dynamic TDMA protocol for wireless sensor networks. *IEEE "GLOBECOM" 2009 proceedings.*
15. Khatarkar, S., & Kamble, R. (2013). Wireless sensor network MAC protocol: SMAC & TMAC. *Indian Journal of Computer Science and Engineering (IJCSE).*, ISSN 0976-5166, *4*(4).
16. Swapna Kumar, S., Nanda Kumar, M., Sheeba, V. S., & Kashwan, K. R. (2012). Power efficient dynamic MAC protocol (D-MAC) for wireless sensor networks. *Journal of Information & Computational Science, 9*(7), 1795–1805.

Predictive Crime Analytics Using Data Science (India and the USA)

Preeti Rachel Jasper and G. Chemmalar Selvi

Abstract Predictive analysis is concerned with the branch of data science used to predict future patterns and trends. This modelling technique can be used to aid society. In recent years, crime against women has skyrocketed, and understanding past history can help to come up with insight that describe the current state of crime and assault in these countries. This research aims to foresee the crime patterns against women in India and the USA. The studies were carried out in both countries to better understand the economic bearing, if any, in crimes committed against women. The data of the past years is studied using extensive EDA (Exploratory Data Analysis) techniques to help understand the problems women face. The data is then normalized, and Linear Regression is executed to predict future trends in crime rates. K-means is then used to create clusters of states with the highest crime rates against women in India. This chapter aims to help women and law enforcement by using technological advancements in the area of data science to predict future trends.

Keywords Data analytics · Crime analysis · K-means · Linear regression

1 Introduction

According to the National Crime Records Bureau of India, a crime against women is committed every 3 min, and the rate of crime shows an increase of approximately 50% for the years 2011–2015. However, data studies help to show that crime against women per capita in India are low compared to other countries. This new-found information helps to understand that the increased rate of crime in India might be directly related to the population growth.

P. R. Jasper (✉) · G. Chemmalar Selvi (✉)
VIT Vellore, Vellore, Tamil Nadu, India
e-mail: preetirachel.jasper2017@vitstudent.com

A. Haldorai et al. (eds.), *3rd EAI International Conference on Big Data Innovation for Sustainable Cognitive Computing*, EAI/Springer Innovations in Communication and Computing, https://doi.org/10.1007/978-3-030-78750-9_13

India has experienced many crimes in the past decade. The datasets of past years are studied using extensive EDA (Exploratory Data Analysis) techniques that aid in understanding the problems Indian women and girls face. The usage of unsupervised learning techniques in this case provides the reader with possible clusters into which future crime may fall into. Predictive Modelling by Feng M provides insights by using least-squares simple Linear Regression problem [1, 2, 23]. Comparative study of various classification algorithms [3, 24] has fuelled the motivation to consider the accuracy provided by supervised learning techniques such as Linear Regression in predicting future crime rates in India compared to other countries, without limiting the study to just one state. The data will then be cleaned and supervised learning techniques such as Linear Regression and unsupervised technique such as K-means can be executed to predict future trends in crime rates with time.

The studies carried out in the USA and India are compared briefly to better understand the economic bearing, if any, in crimes committed against women. The datasets obtained will first be analysed using various methods of exploratory data analysis and we will arrive at a conclusion that will help us carry out linear regression for the same. Both the countries will then be compared according to the results we obtain. We will also perform K-means on India's data to find out the safest and the most dangerous among our states for the women of India.

The research aims to help the women of India by using technological advancements in the area of data science to predict the future state of the country [2]. The datasets obtained from the USA will first be analysed along with India using various methods of exploratory data analysis and aimed to draw certain findings that will help to carry out linear regression for the same. K-Means algorithm will be performed on India to pinpoint the hotspots of crime against women as it is an easy and efficient algorithm that provides near-accurate results [4, 6]. This can help law enforcement agencies focus on these areas more and find remedies to provide a more secure environment.

2 Literature Review

There has been an enormous increase in crime in the recent past [5, 7, 8]. Latest developments in 'Predictive analytics Using Data Clustering Techniques' discussed by A. Anitha [3] showcases the comparative study of various clustering algorithms used to predict the rate of crime. The author states that taking the population of the country into consideration while performing predictive analysis will provide more accurate results. The study concluded that more women were coming forward with complaints in the present years, which then led to a possible increase in the number of filed complaints in the past few years. Both studies show an increase in reported cases from the past decade. 'The crime against women (CAW)' studies done by S Lavanayaa [6] shows the increase in number of reports of rape in India following the Delhi rape case which received national and international attention. Research done by author Sangani showcases the importance of K-means algorithm in creating effective clusters [7].

In 2018, 'Crime rate prediction using data clustering algorithms' [8, 25] conducted a comparative study using K-means and Fuzzy C clustering techniques on unstructured data. The research proposes in-depth studies focussing on smaller and static datasets that can help the user narrow down more accurate predictions. The impact of the individual factors on the results were checked by author Gupta M. for the overall crime rate in Delhi using K-means clustering to classify the respondents or cases into clusters on the basis of degree of crime rate for various factors influencing the crime against women [9, 10, 11]. The author claims of the accuracy provided by K-means. The effective crime prediction and analysis done by Kiani R. uses K-means as a method to classify data based on the level of vulnerability of the area. This is adopted by the paper to generate efficient and clear clusters of hotspots in the country.

The biggest challenge facing the law enforcement is how to efficiently and accurately analyse the increasing volumes of crime data [1]. Author Rishabh Singh performed a variety of unsupervised learning techniques and finally came to the conclusion that K-means is an easy and efficient method for grouping large datasets of crime data. In factors affecting crime against women [10, 12, 17], Kaur states many environmental factors play into the rates of crimes against women in each state and country. The weighted variables are taken into consideration based on PCA and correlation by this paper to create better results.

3 Proposed Methodology

For the study, the following steps have been adopted:

(a) The first step is to identify the problem. In this case, the aim is to understand the rape statistics in a third world vs. a first-world country and hence data pertaining to that is collected.
(b) Exploratory data analysis is conducted to understand the data and to visualize the data for easier and better understanding.
(c) Linear regression is conducted on India and the USA to predict crime trends in each state. This helps to understand the statistical trends in both countries.
(d) K-means is conducted to understand the most dangerous states in India so that law enforcement can focus on them based on the clusters formed.

As shown above in Fig. 1, exploratory data analysis is carried out, and then Linear Regression and K-means are applied to the datasets to predict the future crime trends and the potential hotspots of crime in India.

For this study, the following datasets have been used. All have been obtained from the Open Government Data (OGD) Platform India (data.gov.in).

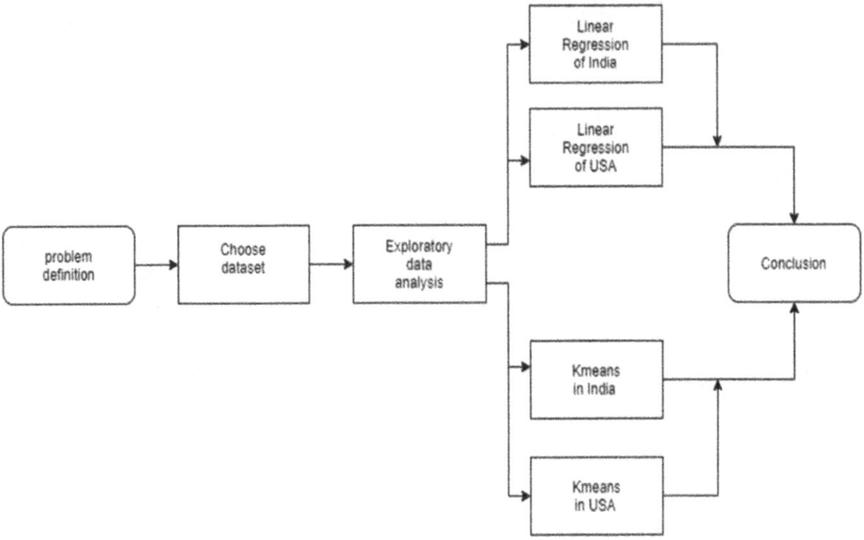

Fig. 1 Block diagram of methodology

1. Crime against Women (2001–2012)
2. World crime against women statistics
3. District-wise crimes committed against Women (2001–2012)
4. Age and sex-wise persons arrested under crime against women
5. Persons arrested under crime against women during 2001–2012
6. The USA rape statistics 2000–2015

Dataset 1 is contributed by Ministry of Home Affairs, Department of States, and National Crime Records Bureau (NCRB). Datasets 3 and 5 are contributed by Ministry of Home Affairs, Department of States, and National Crime Records Bureau (NCRB).

4 Simulation and Analysis

Crime data analysis has a promising future for increasing the effectiveness and efficiency of criminal and intelligence analysis [10, 13, 14, 15, 16]. Crime against women is increasing every year, and as per the research, they have doubled over the past 10 years, according to latest data released by the NCRB (National Crime Records Bureau) [9]. Crime against women over the decade has increased significantly in India as shown in Fig. 2. There has been an increase from ~16000 to ~22000 from 2001–2010.

However, it can be seen that in the USA, the rape rates shows a gradual decrease over the decade. The number of rapes in the USA varies between 100,000 to 80,000

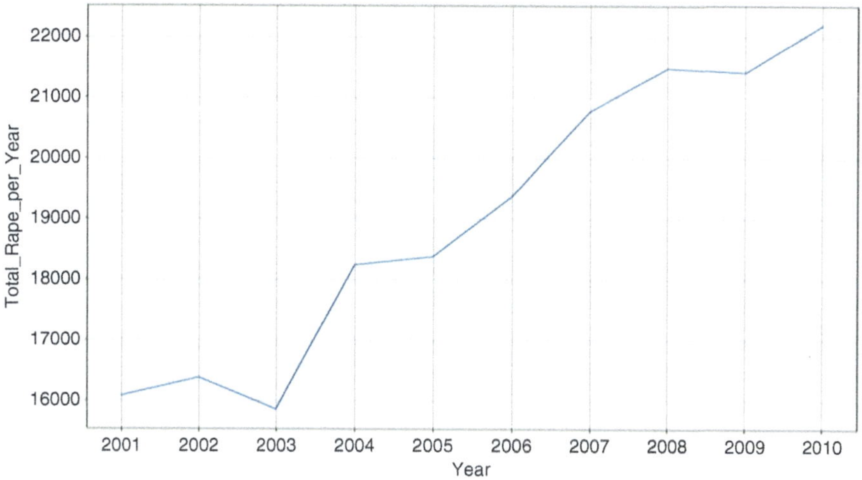

Fig. 2 Pattern of rapes in India over a decade

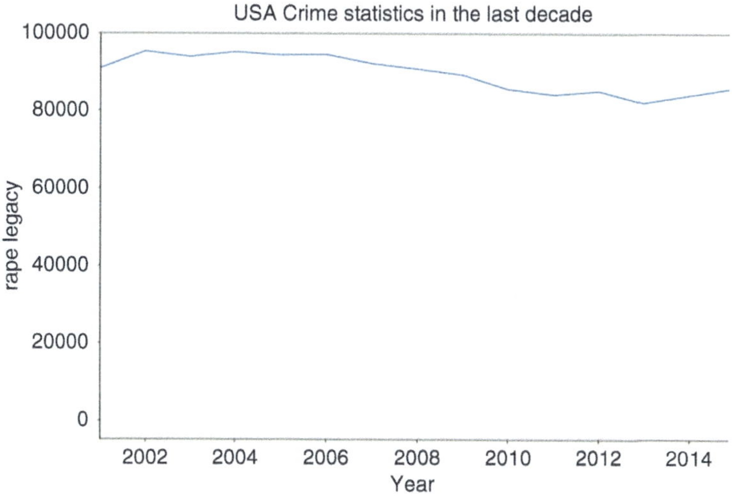

Fig. 3 Pattern of rapes in the USA over a decade

over the decade, which is much higher than in India. Figure 2 shows the pattern of rapes over the decade, from around 16,000 in the year 2001 to about 22,000 in 2010. Crime against women has shown immense growth.

From Fig. 3, it is noticeable that crime in the USA has steadily and slowly declined over the 10 years. The rate of crime against women in the USA is, however, much higher compared to India.

However, data studies show that crime against women per capita in India is low compared to other countries as seen in Fig. 4. This new-found information helps to

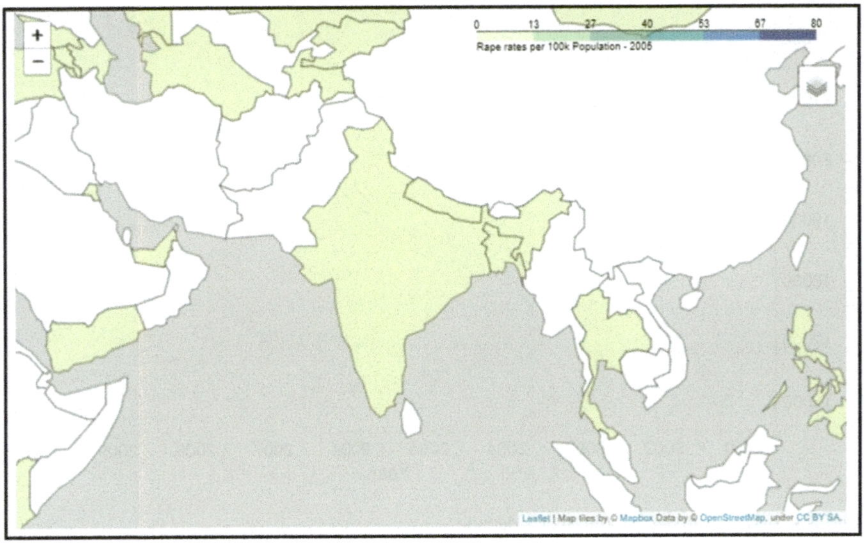

Fig. 4 Folium generated map of India crime rate against women per capita

understand that the increased rate of crime in India might be directly related to the population growth India has experienced in the past decade.

To understand future trends, linear regression was performed to form predicted values for the crimes committed against women over time (Fig. 5).

The calculated predicted value of the average value of rapes per state in India is of value ~34.7. From Fig. 6, a slight increase in the slope can be clearly seen.

Applying the equation for Linear Regression, predicted value of rapes per district is 34.709 per state for a single year in India. Moving on to crimes against women in the USA, it can be seen that the USA has one of the highest numbers of rapes in the world per capita as seen on Fig. 7.

Figure 8 shows the actual vs. predicted values obtained for each year per jurisdiction in the USA.

The regression clearly shows that there has been a decrease in the number of rapes in the USA over the decade. In India, the study has seen a steady rise. The regression line is now plotted based on the predicted values, as seen in Figs. 8 and 9.

The calculated predicted value of the average value of rapes per jurisdiction in the USA is of value 1537.18 for a single year. This value has been calculated through predicted values acquired through linear regression.

This chapter also shows the various characteristics that affect the conditions of rape in India and does an in-depth study of the comparisons of both the countries, India and the USA. India is known as a third-world country and is said to be a 'developing country'. The USA is considered to be a first-world country and economically developed. However, we see that even though India is considered to be a third-world country, women are much safer. A recent study showed that 50% of

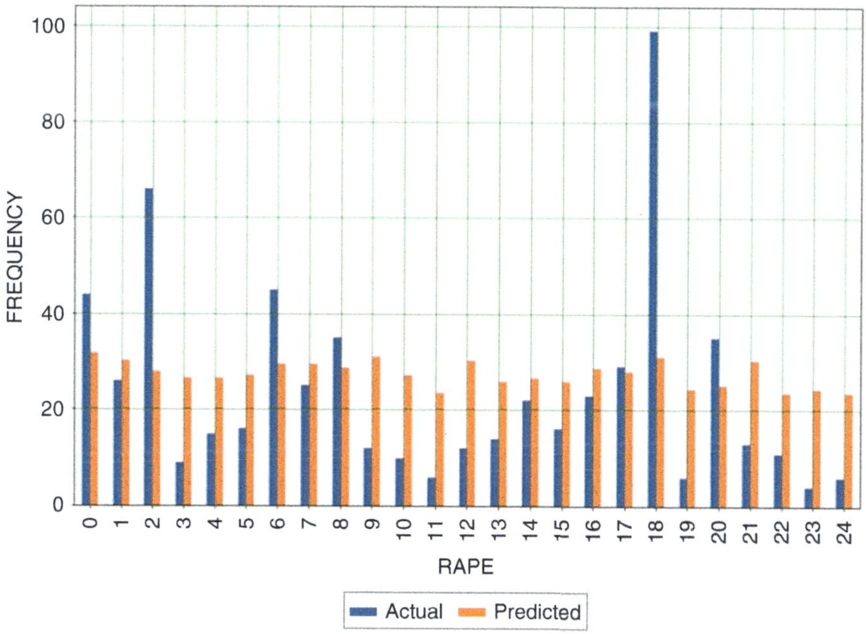

Fig. 5 Actual vs. predicted rates of rape per state in India

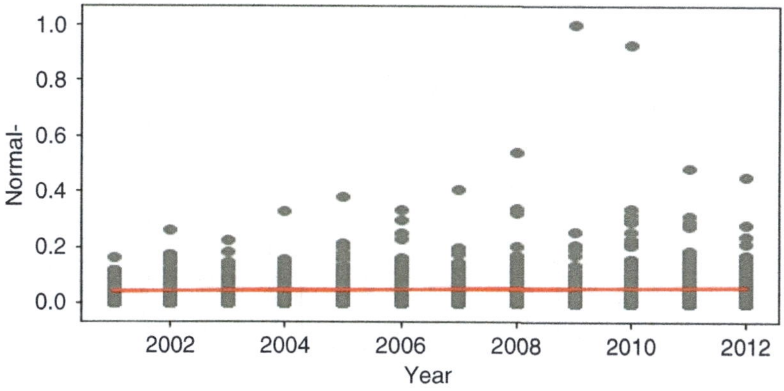

Fig. 6 Normalized linear regression plot of India

rapes go unreported while in the USA 10% only get reported. So even if the amount of rapes was doubled, you can see that India is considerably safer than the USA. This can be because of a number of reasons.

In the predictions, one can see that the y intercept for the USA has a value of 67881.2 and the slope has a value of −32.9087. When applied to the equation, the value of the dependable variable y is found. This will help predict the rape rates per state for the future years. This can help to understand and better prepare for the

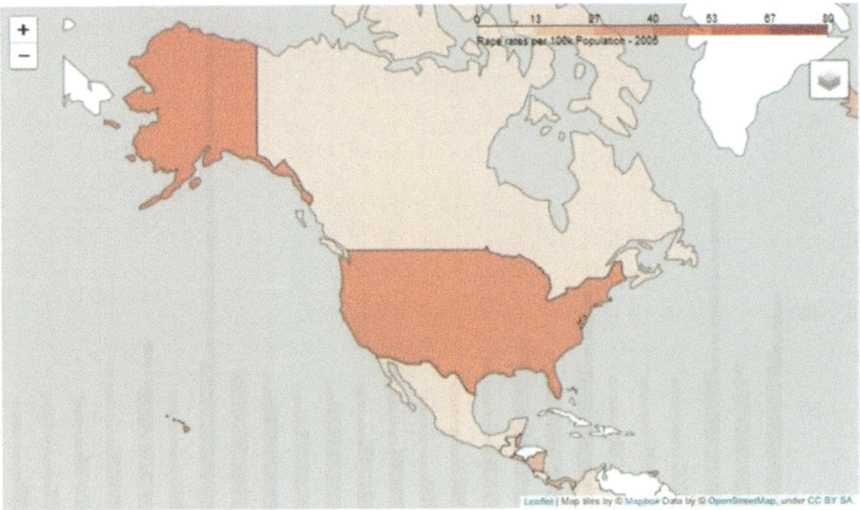

Fig. 7 Folium generated map of the USA crime rate against women per capita

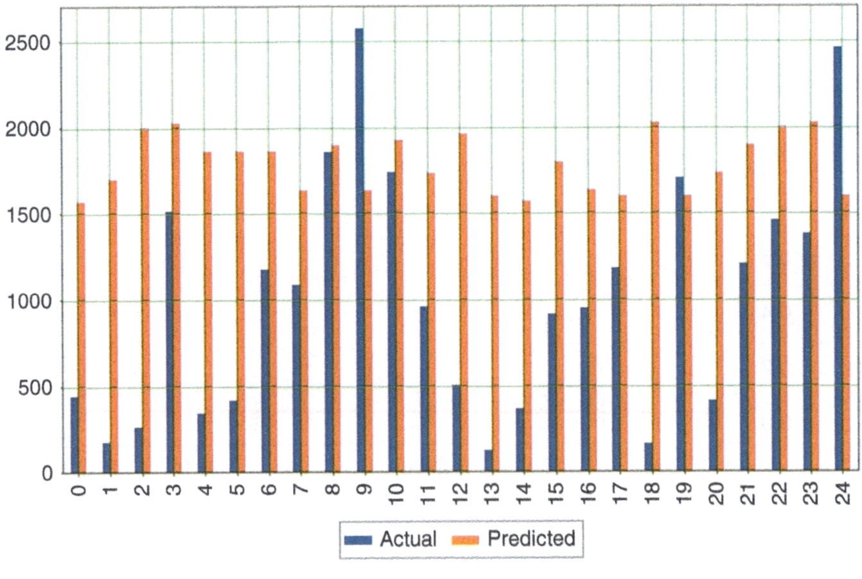

Fig. 8 Actual vs. predicted rates of rape per jurisdiction in the USA

future law enforcement. The predicted value of the average value of rapes per state in the USA is found to be the value of 1537.18 per state for a single year. From the calculations made for India, one can see that the y intercept has a value of −2.585, and the slope has a value of 0.0001. When we apply this to the equation, we find the value of the dependable variable y. Now, the predicted value of the average value of

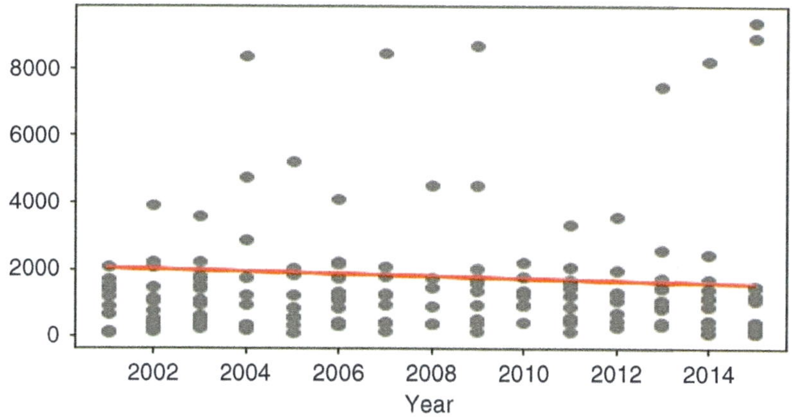

Fig. 9 Linear regression plot of the USA

Table 1 Correlation between variables

	Rape	Cruelty by husband or his relatives	Insult to modesty of women	Assault on women with intent to outrage modesty
Rape	1.000000	0.783092	0.482839	0.932959
Cruelty by husband or his relatives	0.783092	1.000000	0.588474	0.802695
Insult to modesty of women	0.482839	0.588474	1.000000	0.659440
Assault on women with intent to outrage modesty	0.932959	0.802695	0.659440	1.000000

rapes per state in India is calculated. Once, this is done, the value of the predicted variable is 34.7 per state for a single year.

India has a rape rate per capita of 1.8 and is ranked 46th in the world among crimes of rape committed. The USA has a rape rate per capita of 27.3 and is ranked 9th (Table 1).

Now, it can be noticed that the increase in rape statistics in India can be contributed by a few number of states which have very high crime rates against women as opposed to the USA. Finding the hotspots will help law enforcement narrow down on these states, and hence, help in reducing the overall crime rates in India. It can be seen that the highest correlation exists [7, 19, 21] between rapes and assaults against a woman with the intent to outrage her modesty. This can lead one to believe that rape is conducted with the intent to shame the women or to cause question of her modesty. From the world map, it can be noticed that the number of rapes per capita in India is quite low. This is very contradictory to popular belief that women are not at all safe in India. However, this can also be due to a number of reasons, the main one being that the many number of rapes do not get justice

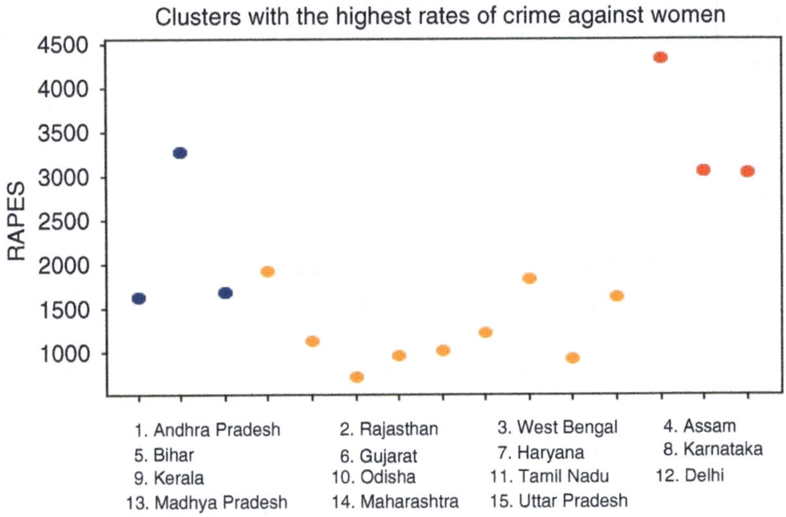

Fig. 10 K-means clusters in India

or in many cases, do not even get reported in India due to the various cultural and religious constraints and judgements cast on a woman subject to such abuse.

The K-means algorithm clusters the given data stored and predicts the possible result. K-mean algorithm is accurate as compared to other clustering algorithms and is simple to understand [7]. K-means is performed on the dataset of India, leading to the formation of three clusters with the highest number of crimes against women. Although the majority of the states in India fall under cluster 0 with low crime rates, these three clusters can be distinguished from the rest of them.

Figure 10 shows the three clusters formed in India. The clusters provide states with various intensities of crime and, hence, can be used to pinpoint the potential hotspots of crime in India. The narrowed down states based on the main variables used for grouping the states into clustering are provided in the tables below. The first cluster contains the states Andhra Pradesh, Rajasthan and West Bengal, as seen in Table 2.

It can be seen in Table 4 that the cluster of hotspots in India that is the most dangerous for women consists of the following states: Madhya Pradesh, Maharashtra, and Uttar Pradesh. The second cluster consists of the following states, namely, Assam, Bihar, Gujarat, Haryana, Karnataka, Kerala, Odisha, Tamil Nadu, Delhi as seen on Table 3.

The Tables 2, 3 and 4 clearly show the states within a cluster with various intensities of crime. By understanding the potential hotspots of crime, law enforcement can narrow down on the problem areas and reduce the overall rate of crimes in India.

Table 2 Cluster one of hotspots

	State/UT	District	Year	Murder	Attempt to murder	Culpable homicide not amounting to murder	Rape	Custodial rape	Other rape	Kidnapping & abduction	…	Hurt/grevious hurt	Dowry deaths	Assault on women with intent to outrage her modesty
33	Andhra Pradesh	ZZ Total	2013	2484	2249	170	1635	0	1635	2119	…	60488	492	6930
611	Rajasthan	ZZ Total	2013	1573	1662	86	3285	0	3285	4986	…	18755	453	4829
787	West Bengal	ZZ Total	2013	2264	3523	415	1685	0	1685	4573	…	18159	481	4913

Table 3 Cluster two of hotspots

	State/UT	District	Year	Murder	Attempt to murder	Culpable homicide not amounting to murder	Rape	Custodial rape	Other rape	Kidnapping & abduction	...	Hurt/grevious hurt	Dowry deaths	Assault on women with intent to outrage her modesty
86	Assam	ZZ Total	2013	1354	579	54	1937	0	1937	4801	...	8171	170	2409
131	Bihar	ZZ Total	2013	3441	3889	259	1128	0	1128	5570	...	35675	1182	331
198	Gujarat	ZZ Total	2013	1118	603	49	732	0	732	2666	...	9917	29	1243
223	Haryana	ZZ Total	2013	982	781	55	971	0	971	2772	...	3171	263	1560
333	Karnataka	ZZ Total	2013	1601	1984	92	1030	1	1029	1836	...	1836	...	21118
354	Kerala	ZZ Total	2013	372	603	104	1221	0	1221	252	...	20399	21	4362
538	Odisha	ZZ Total	2013	1454	2173	39	1832	0	1832	2370	...	10743	395	4618
656	Tamil Nadu	ZZ Total	2013	1936	3007	33	923	0	923	1779	...	20984	118	1271
817	Delhi UT	ZZ Total	2013	517	585	65	1636	0	1636	6294	...	1768	144	3515

Table 4 Cluster three of hotspots

	State/UT	District	Year	Murder	Attempt to murder	Culpable homicide not amounting to murder	Rape	Custodial rape	Other rape	Kidnapping & abduction	…	Hurt/grevious hurt	Dowry deaths	On women with intent to outrage her modesty
409	Madhya Pradesh	ZZ Total	2013	2112	2360	117	4335	0	4335	3354	…	35256	776	8252
455	Maharashra	ZZ Total	2013	2512	2428	203	3063	0	3063	2640	…	30796	320	8132
743	Uttar Pradesh	ZZ Total	2013	5047	5259	1268	3050	0	3050	11183	…	12398	2335	7303

5 Conclusion

Linear Regression analysis was used to effectively predict the future trends of crime in India and the USA. Validation techniques of RMSE, MSE were used to confirm the accuracy of the results obtained through the calculations. K-means was then applied to the datasets of India to form clear-cut clusters of the hot spots of crime in India. This helps law enforcement narrow down stringent measures to the hotspots based on the concentration of crime. It is important to note that when the mean value of rape cases in India per state is calculated, it can be seen that over the decade it has been predicted the value of 34.7 per state for a single year. It was also concluded through validation that in the USA 1537.18 crimes against women is committed per state over a single year. India has a rape rate per capita of 1.8 and is ranked 46th in the world among crimes of rape committed. The studies show a rate of 2 rapes per 100,000 people for India compared to 28.6 rapes/100,000 people for the USA. The USA has a rape rate per capita of 27.3 and is ranked 9th, 15 times more than India. From the data, it can be derived that the USA has16 times more rapes than in India. Even if 50% of the rape cases in India are not reported, and all in the USA are reported, they are still 8 times unsafe.

6 Future Works

The chaper sheds light on the state of the safety of women in India and the USA using Linear Regression to predict future trends and K-means to pinpoint the potential crime hotspots in India. However, one cannot narrow down the complexities of the reasons behind rape to the economic and social conditions. This is definitely part of a much greater venture to understand the state of various countries that fall under different economic backgrounds. In the future, one must take into account the various cases that go unreported in the countries for a better assessment. It has been reported that ~50% of the cases in India go unreported and ~12% of cases in the USA are actually reported. More complex algorithms can be applied to provide more accurate findings. Finding the actual datasets can help produce better and accurate predictions. This study provided us with a surprising fact that the number of rapes per capita in the USA is 15 times more than that in India. This brings the question whether feminism, social and economic standards have a large role to play as assumed by the general public. This study should be done on a number of different countries to better understand the possible reasons behind crime against women.

References

1. Shamsuddin, N. H. M., Ali, N. A., & Alwee, R. (2017). An overview on crime prediction methods. In *2017 6th ICT international student project conference (ICT-ISPC), Skudai* (pp. 1–5). IEEE. https://doi.org/10.1109/ICT-ISPC.2017.8075335.
2. Feng, M., Zheng, J., Ren, J., Hussain, A., Li, X., Xi, Y., & Liu, Q. (2019). Big data analytics and mining for effective visualization and trends forecasting of crime data. *IEEE Access, 7*, 106111–106123.
3. Vaidya, O., Mitra, S., Kumbhar, R., Chavan, S., & Patil, M. R. (2018). Crime rate prediction using data clustering algorithms. *International Research Journal of Engineering and Technology (IRJET)*. e-ISSN, 2395–0056.
4. Pande, V., Samant, V., & Nair, S. (2016). Crime detection using data mining. *International Journal of Engineering Research & Technology (IJERT), 5*(01), 2.
5. Gupta, M., Chandra, B., & Gupta, M. P. (2008). Crime data mining for Indian police information system. *Computer Society of India, 40*(1), 388–397.
6. Lavanyaa, S., & Akila, D. (2019). Crime against women (CAW) analysis and prediction in Tamilnadu police using data mining techniques. *International Journal of Recent Technology and Engineering (IJRTE), 7*(5C), 261.
7. Sangani, A., Sampat, C., & Pinjarkar, V. (2019 April). Crime prediction and analysis. In *2nd International Conference on Advances in Science & Technology (ICAST). SSRN Electronic Journal.*
8. Gandhi, P., & Sharma, S. (2018 March). Approach of predictive modeling on crime against women problem. *International Journal of Recent Research Aspects, 5*(1), 284–288. ISSN: 2349–7688.
9. Palmer, P. B., & O'Connell, D. G. (2009). Regression analysis for prediction: Understanding the process. *Cardiopulmonary Physical Therapy Journal, 20*(3), 23.
10. Goel, P., & Yadav, V. (2013). Crime against Indian women–women crime susceptibility indexes (WCSI): A principal component analysis. *OIDA International Journal of Sustainable Development, 6*(7), 93–102.
11. Anitha, A. (2020). Prediction of crime rate using data clustering technique. In *Soft computing for problem solving* (pp. 443–454). Singapore: Springer.
12. Yamuna, S., & Bhuvaneswari, N. S. (2012). Datamining techniques to analyze and predict crimes. *The International Journal of Engineering and Science (IJES), 1*(2), 243–247.
13. Abbott, D. (2014). *Applied predictive analytics: Principles and techniques for the professional data analyst*. John Wiley & Sons.
14. Pepper, J. V. (2008 December). Forecasting crime: A city-level analysis. In *Understanding crime trends: Workshop report* (pp. 177–210). Washington, DC: National Research Council (The National Academies Press.
15. Han, J., & Kambel, M. (2012). *Data mining: Concepts and techniques*. Morang.
16. Kiani, R., Mahdavi, S., & Keshavarzi, A. (2015). Analysis and prediction of crimes by clustering and classification. *International Journal of Advanced Research in Artificial Intelligence, 4*(8), 11–17.
17. Rahmatika, Y., Sediyono, E., & Widodo, C. E. (2020). Implementation of K-means clustering and weighted products in determining crime-prone locations. *Kinetik: Game Technology, Information System, Computer Network, Computing, Electronics, and Control, 5*(3), 3.
18. Sonaqwanev, T., Shaikh, S., Shaikh, S., Shinde, R., & Sayyad, A. (2015). Crime pattern analysis visualization and prediction using data mining. *International Journal of Advance Research and Innovative Ideas in Education, 1*, 681–686.
19. Agarwal, J., Nagpal, R., & Sehgal, R. (2013). Crime analysis using k-means clustering. *International Journal of Computer Applications, 83*(4), 1.
20. Malathi, A., & Baboo, S. S. (2011). An enhanced algorithm to predict a future crime using data mining. *International Journal of Computer Applications, 21*, 1.

21. Kaur, B., Ahuja, L., & Kumar, V. (2018). Factors affecting crime against women using regression and K-means clustering techniques. In *Industry interactive innovations in science, engineering and technology* (pp. 149–162). Singapore: Springer.
22. Bhattacharya, D. (2018 February). Crime detection technique using data mining and K-means. *International Journal of Engineering Research & Technology (IJERT), 7*(2). ISSN: 2278–0181.
23. Hamdy, E., Adl, A., Hassanien, A. E., Hegazy, O., & Kim, T. H. (2015 July). Criminal act detection and identification model. In *2015 seventh international conference on advanced communication and networking (ACN)* (pp. 79–83). IEEE.
24. Vijayarani, S., Suganya, E., & Navya, C. (2020 Jan-Feb). A comprehensive analysis of crime analysis using data mining techniques. *International Journal of Computer Science Engineering (IJCSE), 9*(1), ISSN: 2319–7323.
25. Garg, R., Malik, A., & Raj, G. (2018 January). A comprehensive analysis for crime prediction in Smart City using R programming. In *2018 8th international conference on cloud computing, Data Science & Engineering (confluence)* (pp. 14–15). IEEE.

Reorganizing Virtual Machines as Docker Containers for Efficient Data Centres

N. VasanthaKumari and R. Arulmurugan

Abstract With increase in the use of virtual machines in various fields, there is a need to enhance the balancing of huge workload in data centres. The traditional way of implementing the cloud is usually done with virtual machines. Working of virtualization can be reconsidered with different and enhanced technology, such as containers in data centre. Docker containers are gaining popularity due to its features such as increase in productivity by reducing the number of resources. In this chapter, virtual machines are compared with the containers, and virtualization techniques are examined so that the user can work with respect to the requirements. Unlike a virtual machine, a container does not have another software layer called Hypervisor. Due to these reasons, containerized applications have better performance characteristics than virtual machine-based applications. We will discuss the benefits of containers over hypervisor in containers. Nevertheless, containers execute directly in the kernel of the virtual machine. Docker containers use engine of the docker as an alternative to hypervisor.

Keywords Container · Docker · Virtual machine

1 Introduction

Subsequent paragraphs, however, are indented. Many establishments have already moved/to move to cloud computing services with the growth of Cloud technologies. Cloud computing also permits to share different resources (hardware/software) over

N. VasanthaKumari (✉)
Presidency College, Bangalore, Karnataka, India

R. Arulmurugan
Presidency University, Bangalore, Karnataka, India
e-mail: arulmurugan@presidencyuniversity.in

A. Haldorai et al. (eds.), *3rd EAI International Conference on Big Data Innovation for Sustainable Cognitive Computing*, EAI/Springer Innovations in Communication and Computing, https://doi.org/10.1007/978-3-030-78750-9_14

the network [1]. Traditional method of implementing cloud is by the usage of virtual machines. In recent days, Dockers have got more prominence than virtual machines due to its characteristics. Virtual machines and Dockers have its own importance and features. Dockers are the newest technology in cloud IT, and virtual machines have been used for many years and play an important role in data centre with variety of sizes [2]. Virtualization works where several applications are hosted on a single server, which also has mapping techniques.

Virtual machine (VM) is a computer that is present in other host OS and which provides variety of services to users. With the use of virtual machines, customers can get various services from the providers of Cloud. There are many devices of the system which can also be virtualized. When working on virtual machines, the user works with the assumption of working on physical machines [3]. Virtual machines are fabricated as a layer in the infrastructure by executing programs. Hypervisor is a primary part which is sandwiched between the hardware and OS for running the virtual machines. There are many types of virtualization (at server level/network level, etc.) executed at various levels.

Docker containers also have the same conception of virtual machine except in terms of time and code. Images of the containers become containers only when they run on Docker engine. With only the transformation in infrastructure, containers are providing services to both windows and Linux-based applications.

This chapter is divided into different components/sections. This chapter mainly focuses on comparison of virtual machines and docker containers with respect to benefits, features, etc. Section 2 overviews on limitations of virtual machines. Section 3 on docker containers, benefits, and its different models. Section 4 explains about differences between virtual machines and Dockers followed by conclusion and future work.

2 Background

In this section, we will discuss the parameters that distinguish virtual machines from containers and associate the infrastructure of both and describes why these two technologies is becoming increasingly familiar [4]. Containers and virtual machines jointly have become an important factor in technology, which also ease to deploy the applications.

2.1 Virtual Machines and Containers

Both virtual machines and containers have the same way of segregation of resources and allocation. Virtual machines act similar to physical computers. There is much dissimilarity between a virtual machine and container, but the main difference is

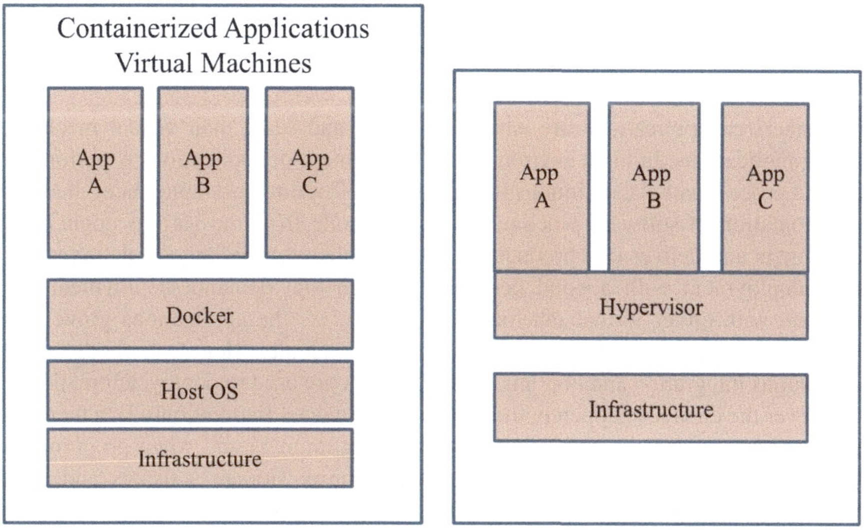

Fig. 1 Container and virtual machines

that containers offer a method to virtualize an OS so that several workloads can execute on one machine. Containers execute with less memory space than virtual machines. Both tools have an impression that single host's task is executed on multiple machines. Both tools are compared with few components like [5] shown in Fig. 1.

Planning of the host machine: In case of containers, kernel is shared between the hosts where the virtual machine has the advantage to execute the kernel which is unlike the machine kernel.

Machine startup: Containers use less resources and the system will start promptly in few seconds. Machine will start in operating system customarily and speed depends on the applications.

Compactness: In virtual machine, there is an issue where applications running on a single host cannot be ported to another host machine, which is allowed in containers. Containers are available in suites that can be used to execute several applications. By nature, containers are lightweight compared to virtual machine. As the containers are lightweight, applications can be easily deployed on server host machines.

Assessment of performance: Virtual machines and containers have its own prominence; hence comparison will not be much successful. Performance is calculated based on many criteria such as speed, cost. Containers use fewer resources and hence it is lightweight. Making a copy is easy in containers compared to virtual machines.

2.2 *Application Deployment in Virtual Machines and Container-Based Environment*

Containerized applications are easier to deploy and scale than virtual machine-based applications. In most environments, deployment of applications is performed using a process called Continuous Integration and Continuous Deployment. It refers to automation of software pipeline to integrate code from the various application developers and deliver into production. Cloud applications in large-scale industries need deployment with a good performance with that demand; deployment can be done with either virtual machines or dockers. As the applications grow, it is important to manage the resources and deploy manually. Hence, methods called continuous integration and continuous deployment are used to deploy automatically whenever the changes happen in the code. In Continuous Integration(CI), a memory is shared for storing the code written by the programmers and when programmer tries to loads the code a snippet from the memory storage activates and does integration of both and executes the test cases. In the market, there are few tools available for performing continuous integration like Jenkins, Bamboo, etc. Git is one of the tools available where developers write the code and merge with the master code [6].

Continuous Delivery (CD) is another tool used to deploy the errors that was fixed, new structures into the server whenever it is needed. Users can do different types of testing that is beyond other types of testing such as integration testing. With these tools, deployment is common which does not have human intrusion. CI/CD tools automates the building of new code into server [7].

3 Related Work

With innovation in research, there are a lot of benefits of containers over virtual machines in terms of security and performance.

Docker containers provide solutions to the concerns with much research by providing lightweight images to ease installation of software [8]. explored that the Docker has the capability to execute multiple containers like Genomic pipelines. According to [9], there are many issues with virtual machines like scheduling, managing the resources which has been determined using containers. In comparing the performance between virtual machines and docker containers, [10] explored that for all the experimentations docker containers indicate request, consume lower execution times compared to virtual machines.

In many researches, more comparisons are done with virtual machines and containers. Virtual machines are delineated to physical machines. It has been considered as a challenge to the researchers to design an effective placement algorithm [11]. The research paper [12] says that dockers can be used in wider sense to reduce traffic in load balancing environments to increase the performance of the data centre. The paper [13] compared Linux containers with virtual machines to improve the performance of systems, which uses the extensions of virtualization.

3.1 Limitations of Virtual Machines

There are many restrictions of using virtual machines. Few of them are [14]:

1. Virtual machines do not provide security as they communicate and share the data; there are chances that the attacker can exploit the system.
2. Overflow of a buffer is common in virtual machines. It has a limited length for the buffer. So, when user attempts to write beyond the limits of the buffer, then that particular condition is called overflow of a buffer.
3. As there is a single hypervisor, then there are chances for failure of the system as that one hypervisor fails then entire system stops working.
4. The main prerequisite for working on virtual machines is the Internet, which basically leads to security issues.

3.2 Docker Containers

Docker is an open-source podium that executes application and makes the program very easy to develop and distribute. Docker provides us an opportunity to deploy the applications into the containers [15]. Docker containers are used for developing and deploying code very fast by reducing the delay, dockers have become a very important tool based on their performance. Users who need the advantage of dockers should have familiarity about Docker Networks, Docker Storage, Docker image, Docker Swarm, etc.

3.2.1 Components of Docker

To understand the architecture of docker, it is required to know about the components used. There are primarily four components inside. They are Docker client and Server, Docker Images and Registries, Docker Containers [16] as shown in Fig. 2.

3.2.2 Docker Client

Docker uses the concepts of client/server architecture in which we know the functionalities of a client and a server. Docker client communicates with the daemon of a docker for constructing or executing of containers. Both communicate using the network interface or APIs. When the command docker run is executed, client directs

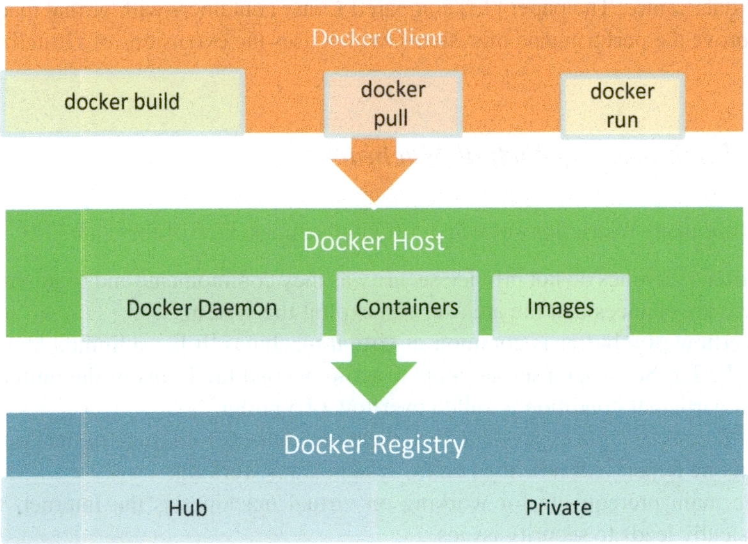

Fig. 2 Architecture of docker container

these commands to dockerd which in turn continues the execution. Docker client can communicate with more than one daemon.

3.2.3 Container

These are the instances of an image. Containers can be created, executed, started or stopped using the API. Containers can be connected to more than one network, store, etc. A container can be explained well by its image and also by the options of arrangement while creating it.

Example: docker run

When this command is executed, docker creates a new network interface to connect to the container. It allows modifying files and directories in the local file system.

3.2.4 Images

There are two ways to build an image. Image is basically a file that is used to run the code in container. When the container executes an image, it turns into instance of that particular container. Images are reusable so that it can be deployed on any machine. Images are comprised of multiple layers. Each layer of docker image can be viewed by a command in Command Line Interface (CLI). There are multiple commands to process docker image such as docker image build, docker image load, docker image pull, docker image push.

3.2.5 Registries

Docker images are reposited in docker registries. Such images can be used to construct images of other applications. Clients communicate with registry using APIs. Pull command is used to reclaim the image from docker daemon. For instance, IBM cloud has container registry, which is used by various customers, individuals, etc. to dispense the images which comprise operating systems, databases, etc.

3.2.6 Advantages of Docker Containers

Containers execute multiple applications. This has many advantages:

1. **Movability and quickness**: They provide the choice of moving applications between the containers very fast and also easily. An application executed on a host can also be deployed and tested into other host machines.
2. **Performance**: It performs well by executing the task very fast where speed is one of the advantages of the containers. Different levels of execution like deploying; testing will be done faster as containers are very small.
3. **Isolation**: A host contains multiple containers and hence applications that need different versions of software can be executed easily.
4. **Container management**: New containers can be created based on the requirement.
5. **Solidity**: Resources are used by the dockers that are accessible more competently as they will not use hypervisor. Comparatively with the virtual machines, docker containers execute on a single host. Docker containers requires less number of resources than machines which are used in the environment of virtual machines.

3.2.7 Shortcomings of Docker Container

Like other technologies, dockers also have drawbacks. There is a need to understand what the requirement for using dockers in the research is. Containers also have disadvantages. They are listed as follows:

1. Containers make use of resources more efficiently than virtual machines. They do not execute on speeds of bare metal.
2. When the containers are closed, the entire data will be vanished unless work is committed.
3. Dockers do not provision Graphical User Interface (GUI) and hence it is not suitable for graphical-based applications.

4. Docker can be executed only on updated machine versions that are only 64 bit. It will not support older versions.
5. Not all applications will be benefited from containers and also do not provide much security.

3.2.8 Setting Up the Environment for Docker and Execute Container

Initially make sure that secure shell (SSH) in loaded into the virtual machine. Start the docker container, and docker daemon provides the permissions to read/write to the Linux socket to users in a group. Docker group should be created, and current user (vagrant) need to be added to the group. Start running the initial container. The docker run command is executed and a name is assigned to the container. If not, docker daemon will assign a name to the container arbitrarily.

dockerd is a continuous action required to manage containers where different programs are used for daemon and client.

vrant@localhost ~]$ ps -ef | grep docker

root 811 1 0 16:47 ? 00:00:01 /usr/libexec/docker/docker-containerd-current –listen unix:///run/containerd.sock –shim /usr/libexec/docker/docker-containerd-shim-current –start-timeout 2mroot 835 1 0 16:47 ? 00:00:02 /usr/bin/**dockerd-current** –add-runtime oci=/usr/libexec/docker/docker-runc-current –default-runtime=oci –authorization-plugin=rhel-push-plugin –containerd /run/containerd.sock –exec-opt native.cgroupdriver=systemd –userland-proxy-path=/usr/libexec/docker/docker-proxy-current –init-path=/usr/libexec/docker/docker -init-current –seccomp-profile=/etc/docker/seccomp.json –selinux-enabled –log-driver=journald –storage-driver overlay2 –add-registry registry.fedoraproject.org –add-registry registry.access.redhat.com

root 1066 1 0 16:47 ? 00:00:00 /usr/libexec/docker/rhel-push-plugin

vagrant 1115 1089 0 16:55 pts/0 00:00:00 grep –color=auto docker

To check with docker running, following command is used:

[vagrant@localhost ~]$docker –version

ok **version 1.13.1**, build b5e3294/1.13.1

To create docker group and start docker Containers:

vrant@localhost ~]$sudo usermod -aG docker $USER

vrant@localhost ~]$ grep docker /etc/group

To run first docker container:

vrant@localhost ~]$ docker run –name hello hello-world

Hello from Docker!

This message shows that your installation appears to be working correctly.

To generate this message, Docker took the following steps:

1. The Docker Daemon will contact Docker client.
2. "hello-world" image will be pulled from the Docker hub by Docker daemon.

3. A new container is produced by the daemon that executes the image and generates the output.
4. Docker client receives the output from daemon, which in turn is directed to the terminal.

The command is run inside the container. Command to run is specified in the image by the image developer when the image is built.

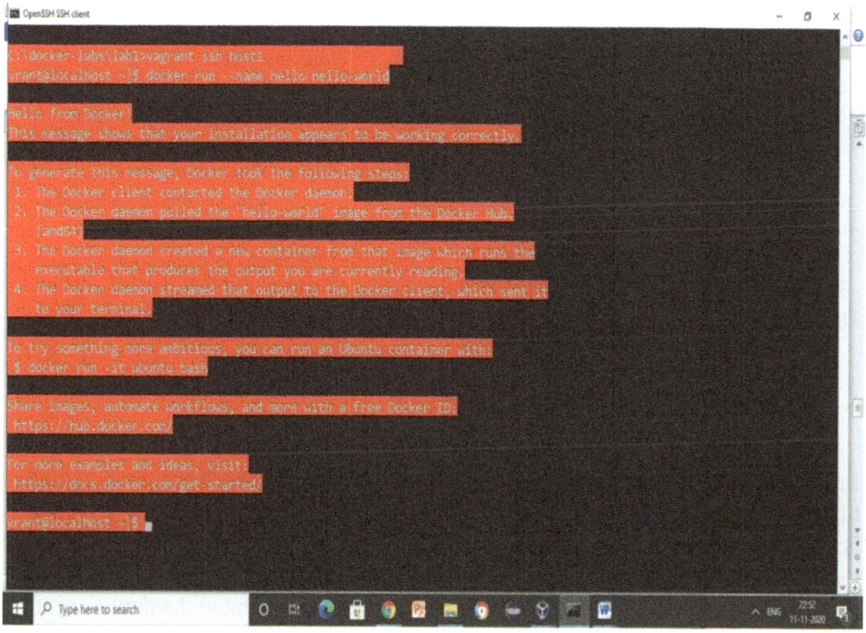

4 Containers in Cloud Environment

Docker Containers are collection of programs which comprise all the necessary components to execute in any type of environment. Containers anticipates and executes irrespective of the environment. Any machine can have different containers with a single operating system which excludes dependencies. Many algorithms were proposed for container scheduling on physical machines. Many technologies exist for understanding the concept of containers. The most used ones are docker, Linux-Vserver.

5 Virtual Machine Versus Containers

Upon different analysis on benefits and limitations of twin technologies, there exist differences based on criteria given in the following Table 1.

Table 1 Comparison of virtual machine containers

Feature	Virtual machine	Containers
Operating system	Need of operating system serving as guest	OS inside can be common among operating system
Portability	A smaller amount	Can be done more
Right to use resources	There is no use of resources as the crow flies	It is possible
Resources required for execution	Requires more in number	Not much required compared to virtual machine
Number of servers required	Requires more servers	Less in number
Need of memory	Less memory is required	Less memory as host operating system is distributed among various applications
Being away from threat	Provides safety to the data inside the host as there is a hypervisor	As the kernel is shared among the applications, safety cannot be assured
Allocation of library files and others	Distribution is not allowed	It can be distributed using different commands of Linux

6 Future Work and Conclusion

In this chapter, two important technologies are compared based on few parameters like portability, security, etc. Both serve same purpose of virtualization. Containers do not require many resources compared to virtual machines. Choosing one of them is the responsibility of a researcher based on the requirement of the research. Continuous Integration and Continuous Delivery tools automate the deployment by integrating the code. In the future, improvements should be carried out by providing more security while using containers. There are different docker models available for variety of Operating Systems. Docker containers are added asset to the current technologies. They are more comfortable than virtual machines in deployment, testing. In future, still various traits of virtual machines and containers need to be discovered. To conclude, dockers perform faster than virtual machines as they use a smaller number of resources and a guest operating system does not exist. Further, there is a contest for all researchers in energy efficiency in cloud and consolidating services with workloads.

References

1. Ala'Anzy, M., & Othman, M. (2019). Load balancing and server consolidation in cloud computing environments: A meta-study. *IEEE Access, 7*, 141868–141887. https://doi.org/10.1109/access.2019.2944420.
2. Singh, S., & Singh, N. (2016). Containers & docker: Emerging roles & future of cloud technology. In *2016 2nd international conference on applied and theoretical computing and communication technology (iCATccT)*. https://doi.org/10.1109/icatcct.2016.7912109.

3. Zhao, H., Zheng, Q., Zhang, W., Chen, Y., & Huang, Y. (2015, December). Virtual machine placement based on the VM performance models in cloud. In *2015 IEEE 34th international performance computing and communications conference (IPCCC)*. https://doi.org/10.1109/pccc.2015.7410296.
4. Yadav, R. R., Sousa, E. T. G., & Callout, G. R. A. (2018). Performance comparison between virtual machines and Docker containers. *IEEE Latin America Transactions, 16*(8), 2282–2288. https://doi.org/10.1109/tla.2018.8528247.
5. Guan, X., Wan, X., Choi, B.-Y., Song, S., & Zhu, J. (2017). Application oriented dynamic resource allocation for data centers using Docker containers. *IEEE Communications Letters, 21*(3), 504–507. https://doi.org/10.1109/lcomm.2016.2644658.
6. Beloglazov, A., & Buyya, R. (2011). Optimal online deterministic algorithms and adaptive heuristics for energy and performance efficient dynamic consolidation of virtual machines in cloud data centers. *Concurrency and Computation: Practice and Experience, 24*(13), 1397–1420. https://doi.org/10.1002/cpe.1867.
7. Sharma, O., & Saini, H. (2017). SLA and performance efficient heuristics for virtual machines placement in cloud data centers. *International Journal of Grid and High Performance Computing, 9*(3), 17–33. https://doi.org/10.4018/ijghpc.2017070102.
8. Gao, Y., Lin, W., & Zhou, J. (2019). Cost-efficient and quality of experience-aware provisioning of virtual machines for multiplayer cloud gaming in geographically distributed data centers. *IEEE Access, 7*, 142574–142585. https://doi.org/10.1109/access.2019.2944405.
9. Arianyan, E., Taheri, H., & Sharifian, S. (2015). Novel energy and SLA efficient resource management heuristics for consolidation of virtual machines in cloud data centers. *Computers and Electrical Engineering, 47*, 222–240. https://doi.org/10.1016/j.compeleceng.2015.05.006.
10. Rasouli, N., Razavi, R., & Faragardi, H. R. (2020). EPBLA: Energy-efficient consolidation of virtual machines using learning automata in cloud data centers. *Cluster Computing, 23*(4), 3013–3027. https://doi.org/10.1007/s10586-020-03066-6.
11. Soltanshahi, M., Asemi, R., & Shafiei, N. (2019). Energy-aware virtual machines allocation by Krill Herd algorithm in cloud data centers. *Heliyon, 5*(7), e02066. https://doi.org/10.1016/j.heliyon.2019.e02066.
12. Wang, T., & Hamdi, M. (2016). Presto: Towards efficient online virtual network embedding in virtualized cloud data centers. *Computer Networks, 106*, 196–208. https://doi.org/10.1016/j.comnet.2016.06.036.
13. Santos, E. A., McLean, C., Solinas, C., & Hindle, A. (2018). How does docker affect energy consumption? Evaluating workloads in and out of docker containers. *Journal of Systems and Software, 146*, 14–25. https://doi.org/10.1016/j.jss.2018.07.077.
14. Amoon, M. (2018). A multi criteria-based approach for virtual machines consolidation to save electrical power in cloud data centers. *IEEE Access, 6*, 24110–24117. https://doi.org/10.1109/access.2018.2830183.
15. Wang, B., Chang, X., & Liu, J. (2015). Modeling heterogeneous virtual machines on IaaS data centers. *IEEE Communications Letters, 19*(4), 537–540. https://doi.org/10.1109/lcomm.2015.2403832.
16. Tarahomi, M., & Izadi, M. (2019). Energy efficiency in virtual machines allocation for cloud data centers with Lottery algorithm. *International Journal of Electrical and Computer Engineering (IJECE), 9*(1), 546. https://doi.org/10.11591/ijece.v9i1.pp546-553.

Secured On-Demand Adaptive Routing (SOAR) Protocol for Data Transmission in IoT Environment

P. Deepavathi and C. Mala

Abstract The emerging concept Internet of Things (IoT) has the capability to communicate data among devices throughout the entire world without human intervention. Existing reactive routing-based protocol needs tremendous bandwidth, consumes more processing time and generates large number of control packets for communication. This chapter proposes an efficient Secured On-demand Adaptive Routing (SOAR) protocol for IoT environment with secured data transmission among nodes. The proposed protocol is simulated and tested by using NS2 simulators. The simulation results show that throughput and packet delivery ratios are better than existing protocols.

Keywords Adaptive routing · Secured transmission in IoT · Routing in IoT · IoT

1 Introduction

Internet of Things plays an essential position in developing smart e-health, business, agriculture, home and learning, etc. Professional applications of IoTs are industrial automation, supply chain management, logistics handling, smart transportation, remote sensing and monitoring applications. Especially, Industry 4.0 brings lot of innovations in IoT field, for example, this industry 4.0 is manless automation and data exchanges in manufacturing technologies. It brings smartness in all areas, for example, smart city, smart home, smart office.

Internet of Things is activated with lot of sensors; all sensors are connected with each other. Each sensor collects information and passes to the next one through a lot of intermediate nodes. These nodes use limited power and dynamic network environment. Here, all the nodes are connected through wireless links. Whether the

P. Deepavathi (✉) · C. Mala
Department of Computer Science and Engineering, National Institute of Technology, Tiruchirapalli, Tamil Nadu, India

A. Haldorai et al. (eds.), *3rd EAI International Conference on Big Data Innovation for Sustainable Cognitive Computing*, EAI/Springer Innovations in Communication and Computing, https://doi.org/10.1007/978-3-030-78750-9_15

communication may be wired or wireless, the communications should be secured between the IoT nodes. In the IoT network, routing plays an important role; the major problems in routing are route breaking, traffic, packet loss, and breaching data. In this chapter, the proposed mechanism focuses on secure data transmission between the sender and the receiver.

Some of the security goals that should be achieved during transmission for avoiding attacks on data and network resources are discussed [17]. The secured routing is very important to make secure communication between devices; to enable secure communication, various types of secured routing protocols are proposed. The routing protocols are configuring routing information and identifying the available routes and creating the routing table. The purpose of routing protocol is exchanging routing information among routing devices. Two categories of protocols are used in IoT routing environment [17].

1.1 Proactive-Based

These protocols follow the table-driven approach, where all the routing information is entered in the routing table, which is available in each router in the network. Routing table in this method is fixed; when any changes occur in the neighbour router, that routing table will be modified. Examples are: Optimized Link State Routing Protocol, Destination Sequenced Distance Vector Routing Protocol.

1.2 Reactive-Based

These protocols follow the on-demand based approach, which is the route to be discovered when needed. Therefore, no fixed routing table is in the router; the router will take decision based on the requirements. Example: Ad hoc On-demand Distance Vector and Dynamic Source Routing protocols.

The high- and low-rate attacks affecting the requested services in connection-less environment are explained well in [18]. The different types of attacks possible in Ad hoc On-demand Distance Vector Routing protocols are explained well in [17]. The existing Dynamic Source Routing (DSR) protocol is on-demand route request protocol. This protocol will find the path when a source node wants to communicate with some other node. So, maintaining fixed routing table is not needed. But, DSR protocol will not take care of damaged link, and it is not suitable to vast networks. DSR protocol consumes high bandwidth and large number of resources to complete the task. This protocol is suitable for static environment and small networks. The performance of DSR protocol will be decreased when the path length is increased. Various types of incoming traffic also affect the routing where eliminating the traffic to avoid attacks and improving the Quality of Services are discussed well [19].

The remaining portions of this chapter are structured as follows: Sect. 2 gives more detail about related work. The detailed processes of working methodologies are explained in Sect. 3. The Simulation Results and Performance Analysis are given in Sect. 4. Finally, the proposed method concludes with Sect. 5.

2 Related Work

In this part, security problems under routing protocols under Internet of Things (IoT) have been surveyed. A lot of existing methodologies do not satisfy the current requirements and problems in routing protocols. The following section covered the secure routing protocols used in the IoT environment.

Rahul and Sujata have explained about various advantages, challenges in IoT and listed the factors that affect the routing process. CupCarbon simulator was used for the simulation process. But, this process only deals with static routing. The dynamic routing and also the implementations are very difficult [1]. Sathya et al. proposed Real Time Secure Route Analysis (RTSRA) method for providing Mobile Ad hoc Network routing security. This technique considered lot of factors for making security between source and destination. The author used two types of measurements to calculate metric value for counting number of nodes in the route. This method concluded with a single route for improving the Quality of Service. The main drawback is that if a single route fails, no further communication will occur [2].

Ahamed, Sauf proposed Energy Efficient Geographic (EEG) protocol for avoiding region problems and also balancing the energy maintenance of the concerned networks. This method is finding the single hop nearest node knowledge for forwarding data packets. If suppose the nearest node is malicious node, then all nodes will be hacked easily [3]. Michele, Salma et al. proposed Distributed Learning Algorithm for routing configurations; this concept dealt with only local communications among all the nodes in networks [4]. In this paper, the author proposed Secure Multipath Routing Adhoc (SMRA) protocol where two types of approaches are considered for finding the factors that are affecting the performance of routing. But, time efficiency was not considered [5].

Sharma KP et al. proposed signal control and balancing the load of routing. Adaptive load tuning techniques were proposed to achieve good network lifetime, but this method suffered by higher delays during routing process [6]. Kumar V and Kumar A have explained load prevention concepts to achieve good energy level and better throughput, but again this routing mechanism makes higher delays [7]. In this paper, the author has explained Energy-aware and Secure Multi-hop Routing (ESMR) protocol. Here, the author has considered three main phases for splitting network into inner zone and outer zone. So, this proposed technique minimizes the routing overhead. The drawback of this concept is applicable only for WSN and will not be suitable for heterogeneous-based scenarios [8].

S.B.Thigale and R.K.Pandy have explained named data networking-based Cross-layer Attack Resistant Protocol (NCARP) where three basic layers are considered for identifying malicious attackers in VANET. The problem is it is only applicable for Named Data Networking (NDN)-based architectures [9].

Gauya et al. explained S2MCBR scheme for splitting network into different regions, each region has one trust manager for checking the strength of the signal, link capacities and link-related problems and measuring the trust value of different routes. If the trust value calculation has any error, then the security also can be compromised [10]. M.T. Hammi explained about single key cryptography where advanced key sharing approach was used. The single key is easily hacked by unauthorized nodes [11].

Garg et al. proposed fuzzy rule-based concept; this is an extension of AODV for identifying secure route between IoT nodes [12]. B. Hammi et al. explained about the networking based on wireless communications, where spoofing attack frequently occurred [13]. In this paper, the author proposed secure multi-hop routing protocol for establishing secure transactions between IoT devices. This protocol ensures authentication before a new node joins into the group by using multiple layers of authentication; it causes memory overhead to all the nodes in the network [14].

The author proposed trust-aware secure routing framework. This concept is very useful for avoiding selfish attack, on and off attack, collusion and bad mounting attack and conflict activities attack. This method uses two values (0 and 1) for identifying trust. The trust value is 0 means no trust between nodes, trust value 1 means good trust between nodes. But the main drawback in this method is memory overhead and processing time overhead [15].

In this paper, the author explained a lightweight and trust-related routing where Intrusion Detection System (IDS) estimates the trust of one object and computes limited computational resources. Generally two types of disturbances are getting from neighbors that are unreliable neighbors and unreliable observations. Here, the author did not consider the unreliable observation attacks [16].

3 Proposed Work

This proposed Secured On-demand Adaptive Routing (SOAR) protocol belongs to asymmetric cryptography where the hash chain technique is used to check the authentication of packet, and this hash function value is used for passing the updated routing information to the neighbour nodes. If a node is getting any updated routing information, it will check the authentication of each entries of the message.

Digital signature is incorporated with Route Request (RREQ) message. The source node S wants to communicate with Node D. Therefore, Node S transmits Route Request (RREQ) message with Digital Signature to next neighbour node; that node will check the signature, then it forwards to next neighbour node till it reaches the corresponding Destination node D (see Fig. 1).

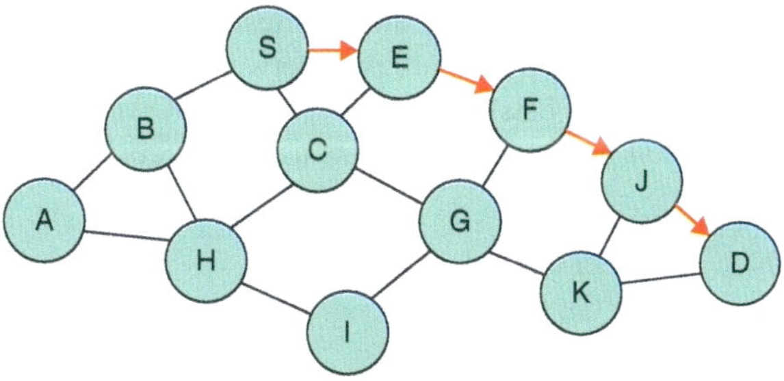

Fig. 1 Route discovery

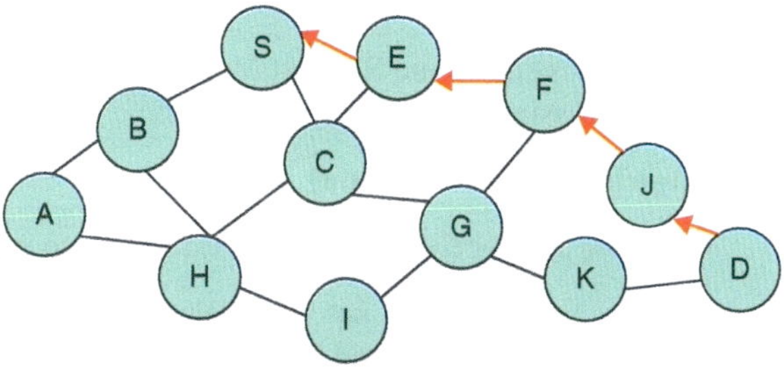

Fig. 2 Route reply

HASHING FUNCTION	MAXIMUM NODE COUNT
BIG HASH	
DIGITAL SIGN	
HASHING VALUE	

Fig. 3 SOAR protocol header

And the corresponding receiver node also will check the signature before producing Route Reply (RREP). The node D sends Route Reply (RREP) message after finding the secure and shortest path (see Fig. 2).

The proposed SOAR protocol header format is shown in Fig. 3. This protocol header has five fields; these fields are doing major functions of checking the authentication and finding secure shortest path for communication among nodes.

In SOAR routing protocol where the hash function is used for ensuring the authentication, it generates the random number that is known as seed. This seed selecting the Maximum Node Count, this maximum node count is used to set the Time To Live value. The Big Hash is used to set the Maximum Node Count Times. While processing RREQ and RREP messages, the hop count hashing with Maximum Node Count checks whether the answer is the same as Big Hash value; if it is same, normal communications continued else the packet is lost. If suppose that link is damaged, the error message will be generated by the nodes (see Fig. 3).

After finds the secure and shortest path, the secure data transmissions will be performed. Here the secure path between S to D is S->E->F->J->D (see Fig. 4).

Algorithm 1 lists the steps involved in SOAR protocol for secure communication among IoT nodes. The notations used in Algorithm 1 are:

N_A = New Node A
N_N = Number of Nodes
G_N = Node Group
S_N = Starting Node
D_N = Receiver Node
I_N = In-Between Node
RREQ = Processing Route Request Message
RREP = Processing Route Reply Message

Algorithm 1: SOAR Routing Algorithm

//Sending and Receiving Route Request and Route Reply

New Node N_A Joins with NN in the Network Group G_N
Group incremented G_N=N_N+N_A
Source SN send RREQ to Destination node DN with DIGITAL SIGN
Intermediate node IN Check(SIGN(SN))
Loop
    I_N forwards RREQ to next neighbor node
    I_N=I_N+1
    D_N Received RREQ
End Loop
D_N Check(SIGN(S_N))
D_N send RREP to S_N

//Calculating Shortest and Secure path

RREQ and RREP Generate Random Number R_N
Repeat
    R_N=set(MaxNodeCount)
    MaxNodeCount setting TTL value
    BigHash=set (MaxNodeCountTimes)
    OutHash= Hash(MaxNodeCount and MaxNodeCountTimes)
    If (BigHash == OutHash)
        Deliver data to the D_N
    Else
        Data Packet will be dropped
    End if
End

CupCarbon simulator used in static routing, but it is difficult to implement it in dynamic routing. Real-time Secure Route Analysis methods provide security for single route, but this method is not suitable for multi-hop routes. Secure Multipath Routing Ad Hoc protocol is used for increasing the performance of routing; here, adaptive load-tuning techniques give good energy level and better throughput,

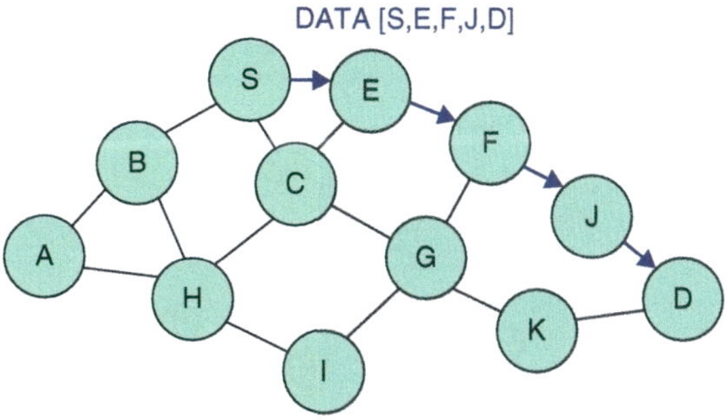

Fig. 4 Secure transmission from source to destination

Table 1 NS2 simulation parameters

Parameters	Values
Runtime	120 s
Coverage region	1000 m * 1000 m
Protocol	DSR
Process interval	0–60s,each step in 20s
Packet load	1000bytes/Packet
Processing speed	200 k
Total nodes	100

but this routing mechanism increased delays. Secure multi-hop routing techniques minimized routing overhead, but this technique is not suitable to heterogeneous networks. Secure multihop routing protocol ensures multiple layers authentication, but it causes memory overhead. These limitations are overcome by this proposed Secured On-Demand Adaptive Routing (SOAR) algorithm. In Dynamic Routing protocol, Hash function is used to provide authentication and Digital Signature is also incorporated in each packet to increase the secure transmission.

4 Simulation Results and Performance Analysis

The Secured On-Demand Ad hoc Routing (SOAR) protocol is simulated by using the NS2 simulator. The following simulation parameters are used for showing the simulation of the proposed protocol (Table 1).

The simulation result of node1 joins in group. Every new node should have proper Digital Signature; after checking the digital signature, it will be added with group G (see Fig. 5).

The performance of sending RREQ message to the neighbor nodes as shown in Fig. 6. In this proposed method, two network groups are doing processes where server1 belongs to one group and server2 is belongs to another group. In the group1 network, lot of nodes are broadcasting route request messages. Each request message is differentiated by different colors.

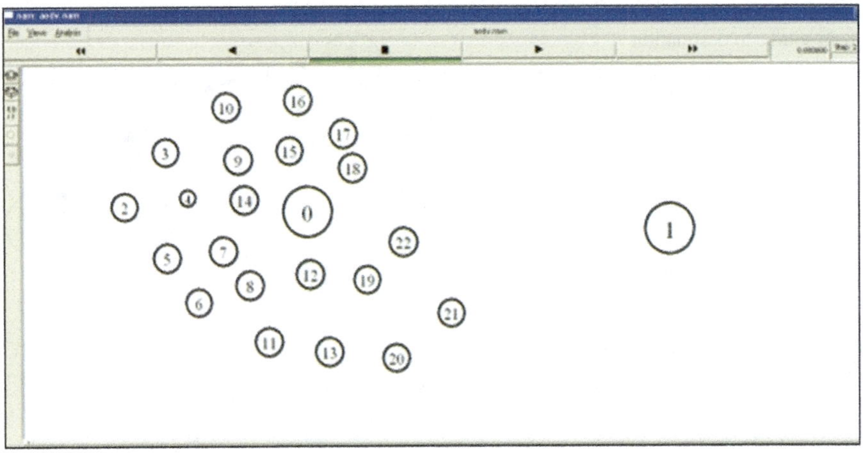

Fig. 5 New node joins

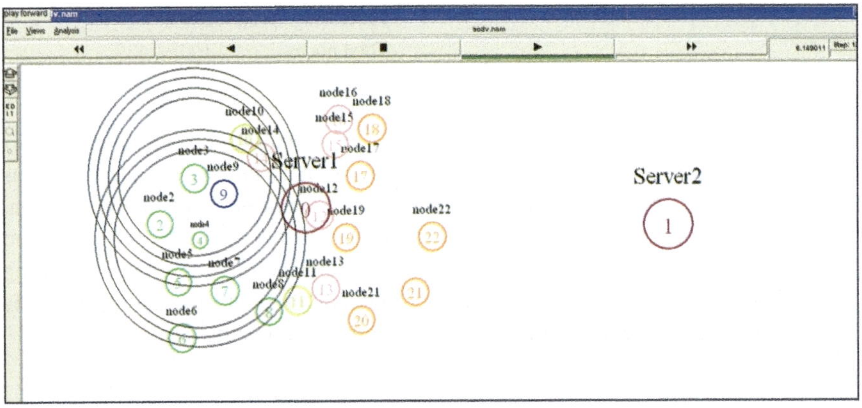

Fig. 6 Broadcasting of RREQ message

The performance of each node is checking the signature of sender's node (see Fig. 7). The authenticated nodes only can send RREP message to the source. Then, the authenticated nodes will be used for further data communication.

The data is securely transmitted to the corresponding receiver node when the protocol finds the secure and shortest path among the nodes (see Fig. 8).

The communication is performed among multiple nodes in NS2 environment; so many numbers of nodes are sharing the data at a time. It shows SOAR protocol increasing the performance of nodes in short time duration (see Fig. 9).

The simulation result of proposed SOAR routing protocol throughput comparisons in NS2 environment where the red colour defines Dynamic Source Routing (DSR) protocol throughput and the green colour defines the SOAR protocol throughput. The comparison takes time in X-axis, throughput in Y-axis. Within the

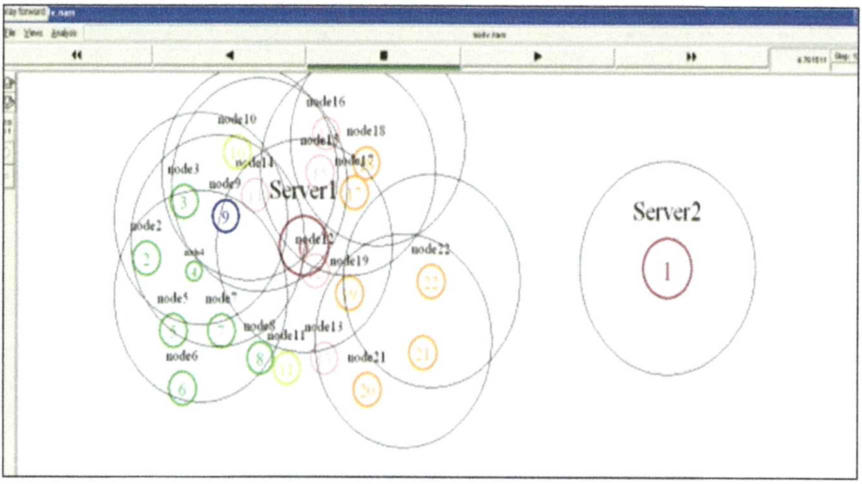

Fig. 7 Verification of digital signature

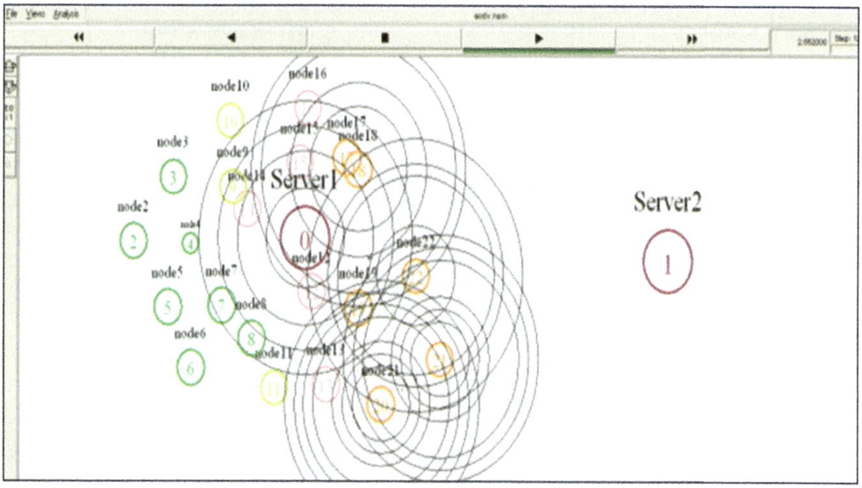

Fig. 8 Transmission of data packets

particular time period, SOAR protocol gives better result than DSR protocol (see Fig. 10).

The simulation result of packet delivery ratio in NS2 environment, where red colour means DSR protocol packet loss and green colour shows SOAR protocol packet loss. For comparing these two protocols, time is chosen for the X-axis and packetloss is chosen for Y-axis. Within the particular time period, the DSR protocol is suffered by a lot of packet loss, but in the SOAR protocol the packet loss is very less. So, SOAR protocol is better than the existing DSR protocol (see Fig. 11).

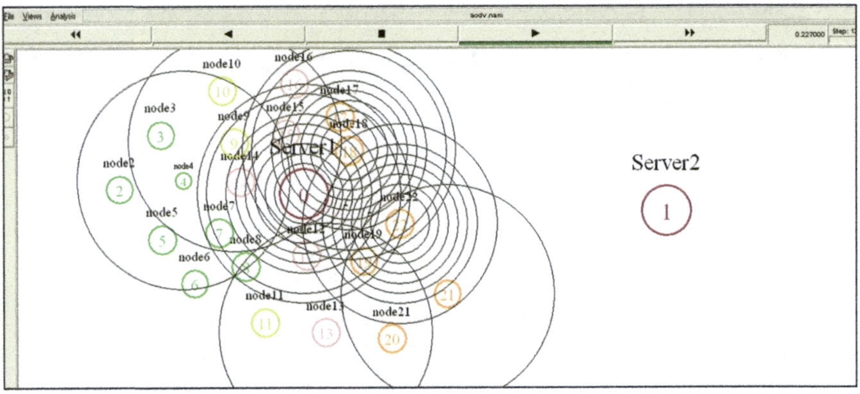

Fig. 9 Communication among multiple nodes

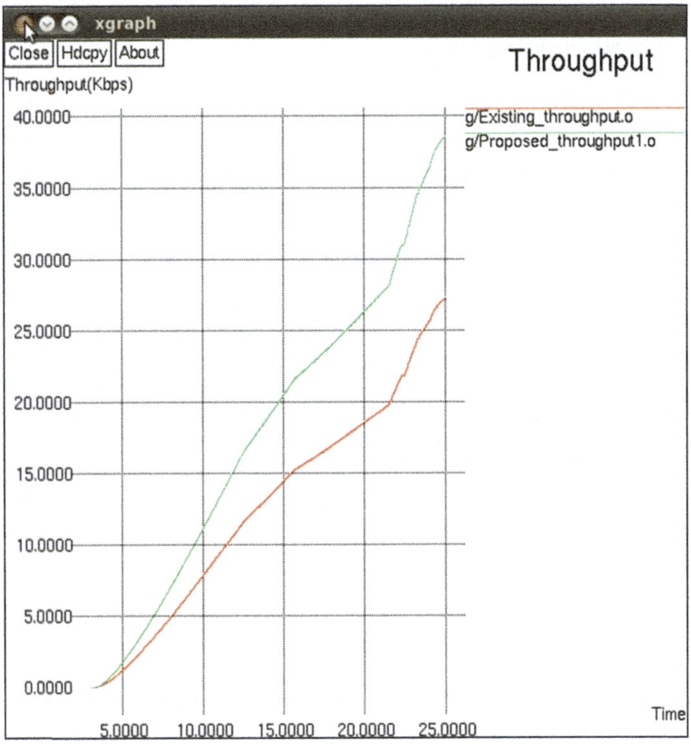

Fig. 10 Time vs throughput

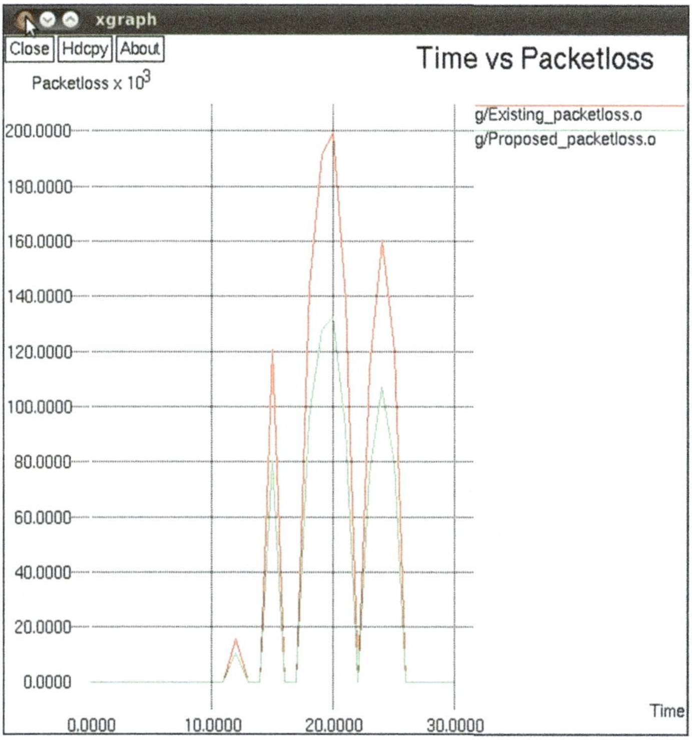

Fig. 11 Time vs packetloss

5 Conclusion

This chapter proposes a Secured On-demand Adaptive Routing (SOAR) protocol for efficient data communication among devices without human intervention. This protocol provides authenticated route request, route reply message and finds shortest secure path for communication among nodes. This protocol is suitable for tremendous networks and consumes less processing time. The NS2 simulation results show that SOAR protocol gives better throughput and minimum packet loss than the existing routing protocol.

References

1. Rahul, S. (1999). Routing in IoT networks using CupCarbon simulator. In *7th international conference on signal processing and integrated networks 2020* (2nd ed.).
2. Sathya, R. (2020). Designing the routing protocol with secured IoT devices and QOS over MANET using trust-based performance evaluation method. *Journal of Ambient Intelligence and Humanized Computing*. https://doi.org/10.1007/s12652-020-02358-4.

3. Ahmad, S., et al. (2020). Towards energy and performance aware geographic routing for IoT-enabled sensor networks. *Computers and Electrical Engineering, 85*, 106643.
4. Michele, S., et al. (2020). Distributed learning algorithms for optimal data routing in IoT networks. *IEEE Transactions on Signal and Information Processing Over Networks, 6*, 179–195.
5. Hammi, B., Zeadally, S., Labiod, H., Khatoun, R., Begriche, Y., & Khi, L. K. (2020). A secure multipath reactive protocol for routing in IoT and HANETS. *Adhoc Networks, 103*, 102118.
6. Sharma, K. P., & Sharma, T. P. (2019). Dynamic load tuning for energy-hole avoidance in Corona model for a wireless sensor networks. *International Journal of Sensor Networks, 30*(1), 56–68.
7. Kumar, V., & Kumar, A. (2019). Improved network lifetime and avoidance of uneven energy consumption using load factor. *Journal of Ambient Intelligence and Humanized Computing, 10*(4), 1425–1432.
8. Haseeb, K., Islam, N., Almorgen, A., & Din, I. U. (2019). Secret sharing-based energy-aware and multi hop routing protocol for IoT based WSAN. *IEEE Access, 7*, 79980–79988.
9. Thigale, S. B., Pandey, R. K., et al. (2019). Light weight novel trust based framework for IoT enabled wireless network communications. *Periodicals of Engineering and Natural Sciences, 7*(3), 1126–1137.
10. Gauya, D. (2019). *A novel dynamic reputation-based source routing protocol for mobile adhoc networks* (Vol. 77). New York: Springer.
11. Hammi, M. T., & Hammi, B. (2018). A Serhrouchni Bubbles of trust: A decentralized blockchain-based authentication system for IoT. *Computers & Security, 78*, 126–142.
12. Hao, S. (2018). A stable and energy-efficient routing algorithm based on learning automata theory for MANET. *Journal of Communications and Networks, 3*(2), 52–66.
13. Hammi, B., & Fayad, A. (2017). IoT technologies for smart cities. *IET Networks, 7*(1), 1–13.
14. Chze, P. L. R., & Leong, K. S. (2014). A secure multi-hop routing for IoT communications: Internet of things. In *(WF-IoT), 2014 IEEE world forum on, IEEE* (pp. 428–432).
15. Raza, S., Voigt, T., et al. (2013). Lithe: Light weight secure coap for the IoT. *IEEE Sensors Journal, 13*(10), 3711–3720.
16. Marchang, N., & Datta, R. (2012). Light-weight trust-based routing protocols for mobile adhoc networks. *IET Information Security, 6*(2), 77–83.
17. Parul, et al. (2010). A comparative study for secure routing in MANET. *International Journal of Computer Applications, 4*(5), 0975–8887.
18. Punitha, V., & Mala, C. (2020). Traffic classification for connectionless services with incremental learning. *Computer Communications, 150*, 185–199.
19. Punitha, V., & Mala, C. (2020). Traffic classification in server farm using supervised learning techniques. *Neural Computing and Applications, 33*, 1279–1296.

Service Measurement Index-Based Cloud Service Selection Using Order Preference by Similarity to Ideal Solution Based on Intuitionistic Fuzzy Values

T. Thasni, C. Kalaiarasan, and K. A. Venkatesh

Abstract Cloud Service Providers are vendors that offer IT services on the Internet. Cloud computing is a concept used to store and access data over Internet. As more and more IT systems are externalized, it has become important for long-term success to ensure that you choose the right cloud providers. Huge number of services are provided by industry leaders such as Microsoft, Amazon and Google to smaller niche players. But, choosing the right cloud provider out of so many is a defined selection and procurement process appropriately weighted towards your unique set of needs. The selection of best cloud service provider is an MCDM approach; several ranking methods such as AHP, TOPSIS etc. have been introduced by researchers. In order to reliably evaluate cloud resources, CSMIC introduced the service measurement index attributes. The aim of this chapter is to compare cloud service providers based on SMI attributes using a ranking approach. The suggested method uses the algorithm called Technique for Order Preference by Similarity to Ideal Solution (TOPSIS), and the uncertainty involved in data is addressed by intuitionistic fuzzy values. The SMI attributes are used as the criteria for the ranking and selection of the best cloud provider.

Keywords CSMIC · MCDM · SMI · Intuitionistic · CSP · TOPSIS

1 Introduction

The next wave of computer architecture is called cloud computing. It is in effect an Internet-enabled confluence of computing resources. Amazon, Google, the sales and Microsoft services are the main commercial cloud computing services. Examples

T. Thasni (✉) · C. Kalaiarasan
Presidency University, Bengaluru, India
e-mail: thasni.t@presidencyuniversity.in; kalaiarasan@presidencyuniversity.in

K. A. Venkatesh
Myanmar Institute of Information Technology, Mandalay, Myanmar
e-mail: ka_venkatesh@miit.edu.mm

A. Haldorai et al. (eds.), *3rd EAI International Conference on Big Data Innovation for Sustainable Cognitive Computing*, EAI/Springer Innovations in Communication and Computing, https://doi.org/10.1007/978-3-030-78750-9_16

of common technology options are applications, platforms and infrastructure. The storage, information and customer data services are focused on the Internet cloud computing. Those that would like to use cloud providers should receive them across the network. There are several cloud services based on their utility and are nearly the same. It is an essential challenge to select the right cloud service. The various authors suggest various cloud services rating strategies. MCDM approaches such as AHP, TOPSIS are the few approaches used in the literature. It is important to select the best Cloud provider among the available ones accurately so that clients and service providers can boost their trust. The SMI (Service Measurement Index) developed by CSMIC (Cloud Service Measurement Initiative Consortium) has developed a set of main cloud service assessment indicators (KPIs) [1]. The cloud provider's selection problem is a multicriteria decision-making problem (MCDM) in which many QoS factors play an important role, in particular determining the best cloud service among choices. Thus, a multicriteria decision-making methodology may be suitable for dealing with consumer or customer requirements and assessing the services provided by the cloud provider as suitable. In this chapter, the authors proposed a ranking and selection approach for selecting the best cloud provider using the TOPSIS method. The SMI attributes like Accountability, Billing, Security, Usability and Performance are used as the criteria for selection of CSPs. The ranking approaches that are called as MCDM techniques such as Fuzzy AHP with Delphi, AHP, Fuzzy ELECTRE (Elimination and Option Expressing Reality)–Fuzzy TOPSIS have been proposed by different researchers in the literature in [2–6], respectively. But in the case of cloud provider dataset, it is impractical to get accurate data. So the uncertainty in the data is handled by intuitionistic fuzzy values in this chapter. This method does not exist in the literature. Based on the following way, the next section of this chapter is organized. Section 2 provides a short description of the works relevant to the cloud service selection. The assessment criteria required for the measurement of cloud providers are specified in Sect. 3. Section 4 presents some basic and fundamental concepts that are essential for our work, and Sect. 5 explains the proposed structure with its main elements and results.

2 Related Work

Service quality (QoS) is the key factor in the selection of a cloud provider according to users' needs. The article [4] addresses the fuzzy Analytical Hierarchy method (Fuzzy-AHP), in order for the selection of cloud service providers to effectively resolve both the quantitative and qualitative factors of preference. When meeting the conditions and criteria of cloud provider selection, the decision maker must select the best option. This issue could be modelled like a problem of decision-making with several parameters (MCDM). A pilot case study was carried out in this research in which the issue of selection of the CC provider was modelled as an MCDM. Multicriteria decision-making techniques for the right cloud service were proposed by Rehman et al. [7] in IaaS, but those multicriteria decision-making strategies are not adequate for fuzzy data. Garg et al. [8] have built the AHP framework for selection of cloud providers. Cloud-Service Metrics-Consortium (CSMIC) QoS criteria were determined by the authors, and main performance indicators (KPIs) were used to evaluate the cloud service. However, the selection and rankings of the same service is solely based on the CSMIC quantifiable parameters and does not take into account the attributes that cannot be quantified in the selection of trustable CSPs from QoS. In this paper, the researcher neglected to consider the fuzziness associated with the assessment criteria for cloud service selection. In [9], researchers identified an Elimination and Selection Expressing Reality (ELECTRE) multicriteria decision-making method that was using user preferences. In [10], authors proposed a mechanism for the selection of cloud services based on multidimensional trust. An evidential reasoning strategy that combines both trust value based on perceptions and trust value based on reputation is based on direct and indirect trust data, respectively, to define trustworthy services. In the form of historical customer feedback scores, multidimensional confidence evidence reinforcing the trustworthiness of cloud services on the different aspects are presented. The ER approach then adds multidimensional confidential valuations in order to achieve the confidence attribute in real time such that certain types of trustworthy cloud services are chosen for active users. R. Ranjan Kumar et al. [11] proposed a hybrid approach that combines AHP and TOPSIS to choose the most suitable service for customers by addressing the significance and importance of better service in the cloud, and had compared the previous methods and identified the drawbacks of those techniques. Five criteria were used by the authors to identify the best cloud services among the many acceptable cloud services. The weights of the parameters allocated in this proposed system by AHP and even TOPSIS are used to evaluate the cloud services. The weights derived using the AHP approach are utilized in the TOPSIS-based selection procedure to determine the rankings of cloud alternatives. Vague assessment criteria were not also taken into account in this paper. In choosing the right cloud service provider, the experience of current customers can also be helpful. The article [12] describes and defines QoS metrics so that consumers and suppliers can both communicate their desires and bid in a quantified way. A versatile, dynamic structure is proposed using the Ranked Voting method that takes inputs and provides the best provider for output requirements from the user. Users have different types of applications on relevant cloud platforms that should be carried out. As a consequence, users could experience problems choosing the best service. Therefore, the choice of a system for comparing services and selecting the best service have been seen as a challenge.

3 Evaluation Criteria

According to the specifications of the International Organization for Standardization (ISO) [1], CSMIC has established SMI attributes. Based on SMI characteristics such as Agility, Service Assurance, Price, Cost, Security, Privacy, Usability and

Accountability, the client will be able to select the best cloud provider that meets the QoS criteria.

3.1 Security and Privacy

The safety of data is one of the key issues about security and privacy in cloud computing. In these clouds, millions of users have stored their important data, which is what makes it more risky to encrypt every bit of information. Data protection has become a major problem in cloud computing, with different data being transmitted to various storage devices and computers, including PCs, servers, and various mobile devices, such as smart phones and wireless sensor networks. The benefits that cloud computing gives us have become an important part of our lives. This is why, if users want to trust the system again, protection and privacy in cloud computing need to take action. To be able to become a part of the cloud computing environment, it is necessary to gain the trust of these users.

3.2 Usability

Usability is the degree to which specified users may use a device, product, or service to achieve specified objectives with performance, effectiveness, and satisfaction in a specified context of use Although users experience the cloud through a user interface, by setting up the cloud service and its features, organizations allow the end-user experience. Via its applications, the end user communicates with the cloud and its cloud-based services. These applications are carried out by cloud and service attributes and will display behaviours that can be tracked back to the cloud. Application testing shows the usability of the cloud and cloud applications that the end user has access.

3.3 Accountability

In the business context, accountability is about complying with measures that give effect to practices articulated in the given guidelines. The complexities of compliance with data security and business regulations are an obstacle to delivering cloud services for the cloud service provider (CSP), and the costs and implications of non-compliance are a serious concern for the cloud service provider (CSP). Accountability will assist us in overcoming these issues in terms of trust and complexity. It is particularly useful to protect sensitive or confidential information, increase customer confidence, clarify the legal situation in cloud computing and promote cloud computing. Accountability offers a way forward in resolving data security issues resulting from the cloud handling of personal data, but these problems surpass the handling of personal data and generalize it to all forms of data, beyond privacy concerns.

3.4 Billing

The aim of the providers is to optimize revenue through their pricing schemes, while the primary objective of the consumers is to have a fair price for quality of services (QoS). Cloud Computing's pricing model is more flexible than traditional models. Each cloud provider has its own pricing mechanism. Cloud Computing is mainly based on fulfilling and achieving Quality of service (QoS) assurance for clients.

3.5 Performance

Performance is one of the main considerations to take into consideration when evaluating a cloud application, since it can have a direct effect on the user experience. Usually, this test practice is conducted to determine such performance characteristics such as output, reactivity, bottleneck, limitations, and latency when the application is under different workloads.

4 Fuzzy TOPSIS

The performance-based scores and the parameter weights are defined by intuitionistic fuzzy values, are calculated in terms of language. Below is the suggested IFTOPSIS form.

4.1 Normalize and Aggregate the Performance Ratings

Let $a_{ijk} = \left(p_{a_{ijs}}, 1 - f_{a_{ijs}} \right), a_{ijk} \geq 0,\ s = 1, 2, \ldots\ldots k, j = 1, 2, \ldots. m, i = 1,$ 2, $\ldots n$, It is the performance rating provided by decision-maker *Dk* for criterion *Cj* to alternative *Ai*.

$a_{ij} = \left(p_{a_{ij}}, f_{a_{ij}} \right)$ of alternative A_i under criterion C_j is the aggregated performance rating which can be evaluated as:

$$a_{ij} = \left(\frac{1}{k} \right) \otimes \left(a_{ij1} + a_{ij2} + a_{ij3} + \ldots a_{ijk} \right) \tag{1}$$

$$a_{ij} = \left(p_{ij}, 1 - f_{ij} \right) \text{ can be normalized as follows :}$$

$$r_{ij} = \left(\frac{p_{aij}}{p^{+}{}_{aij}}, \frac{\left(1 - f_{aij}\right)}{\left(1 - f_{aij}\right)^{+}} \right), \tag{2}$$

$$p^{+}{}_{ij} = \max_{i} p_{ij} \text{ and } \left(1 - f_{aij}\right)^{+} = \max_{i} \left(1 - f_{a_{ij}}\right) \text{ for } j \in B$$

$$r_{ij} = \left(\frac{p^{-}{}_{ij}}{p_{aij}}, \frac{\left(1 - f_{aij}\right)^{-}}{\left(1 - f_{aij}\right)} \right), \tag{3}$$

$$p^{-}{}_{ij} = \min_{i} p_{ij} \text{ and } \left(1 - f_{aij}\right)^{-} = \min_{i} \left(1 - f_{a_{ij}}\right) \text{ for } j \in C$$

4.2 *Normalize by Aggregating the Importance Weights*

Let $wt_{jk} = \left(p_{wt_{js}}, 1 - f_{wt\,js} \right), x_{ijs} \geq 0, j = 1, 2, \ldots .m, s = 1, 2, \ldots .k$, the weight that decision-makers give to criterion *Cj. And* also the aggregated importance weight is given as $wt_{j} = \left(p_{w_{tj}}, 1 - ft_{w\,j} \right)$ of criterion C_j can be calculated as:

$$wt_{j} = \left(\frac{1}{k}\right) \otimes \left(wt_{j1} + wt_{j2} + wt_{j3} + \ldots wt_{jk} \right) \tag{4}$$

And the aggregated weights can be normalized as follows:

$$wt'_{j} = \frac{wt_{j}}{\sum_{j=1}^{m} wt_{j}} = \left(\frac{p_{wtj}}{\sum_{j=1}^{m} p_{wtj}}, \frac{\left(1 - f_{wt_{j}}\right)}{m - \sum_{j=1}^{m} \left(1 - f_{xij}\right)^{+}} \right) \tag{5}$$

where $wt'_{j} is$denoting the normalized wt_j

4.3 Normalized Decision Matrix (NDM) Calculation

The NDM is given as $V_{ij} = [v_{ij}]_{n \times m}$ where $v_{ij} = r_{ij} \otimes wt_j$.

$$v_{ij} = \left[p_{\left(r_{ij \times wt'_j}\right)}, (1-f)_{\left(r_{ij \times wt'_j}\right)} \right] = \left[p_{r_{ij}} \times p_{wt'_j}, \left(1 - f_{rij}\right) \times \left(1 - f_{wt'_j}\right) \right] \quad (6)$$

4.4 Determine Ideal Positive and Ideal Negative Solutions

+ ideal and − ideal solutions are described as:

$IA^+ = (1{,}1), j \in \Omega_b$
$IA^- = (0{,}0), j \in \Omega_c$

And also the distance between each alternative IA^+ and IA^- can be obtained as:

$$dt_i^+ = \sqrt{\frac{1}{2m} \sum_{j=1}^{m} \left[\left(p_{vij} - 1\right)^2 + \left(f_{v_{ij}}\right)^2 \right]}, i = 1, 2 \ldots n \quad (7)$$

$$dt_i^- = \sqrt{\frac{1}{2m} \sum_{j=1}^{m} \left[\left(f_{vij} - 1\right)^2 + \left(p_{v_{ij}}\right)^2 \right]}, j = 1, 2 \ldots .n \quad (8)$$

where dt_i^+ gives the value that represents the separation between IA^+ and each alternative, and the separation between each alternative and IA^- is denoted by dt_i^-.

4.5 Closeness Coefficient Computation

The closeness coefficient may be computed as:

$$Cl_i = \frac{dt_i^-}{dt_i^- + dt_i^+}, \mathrm{i} = 1, 2 \ldots \ldots .n \quad (9)$$

Authors can rank all the alternatives as per the closeness coefficient, and can select the best alternative from them.

5 Ranking and Selection Process

A simple success metric for cloud service providers is the quality of service. In promoting the various services provided by the CSPs, high service quality can have a major impact. The assessment of cloud service providers based on the quality of services has, therefore, become an important issue in the choice of providers. Recently, clients such as business organizations have been interested in assessing the service quality of the providers required to pick the best CSP. The selection of quality of service-based criteria is an important factor in the ranking and selection of cloud service providers. The CSMIC consortium had proposed service measurement index attributes as an important criterion for the ranking and the selection of the Cloud Service providers. The SMI attributes are accountability, usability, performance, billing and security as given in Fig. 1.

Fuzzy TOPSIS is a decision-making approach with several parameters, and the principle of linguistic TOPSIS combined with the intuitionistic fuzzy values addresses the fuzziness involved in the problem. This is the first work according to the literature that uses the TOPSIS method and fuzzy intuitionistic values to compare multiple cloud providers based on SMI attributes. In this work, Intuitionistic Fuzzy Valued Technique for Order Preference by Similarity to Ideal Solution is used for ranking the CSPs based on SMI attributes. The ranking of services based on SMI characteristics such as usability, performance, billing, security and accountability performed in this work is not yet published in the literature. Figure 2 shows the steps in the evaluation and selection of the providers.

5.1 Steps Included in Suggested Approach

The ranking of the service provider is done using the Fuzzy TOPSIS method. The detailed step-by-step process involved in the ranking and selection of cloud service provider is shown in Fig. 2.

As mentioned above, the cloud service providers are ranked by IFTOPSIS. Service Measurement Index characteristics are the parameters used for rating and choosing the relevant cloud service providers in this chapter. The dataset of indi-

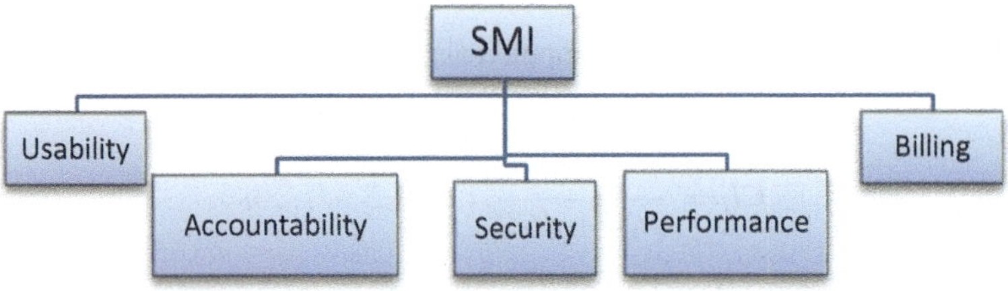

Fig. 1 The SMI attributes

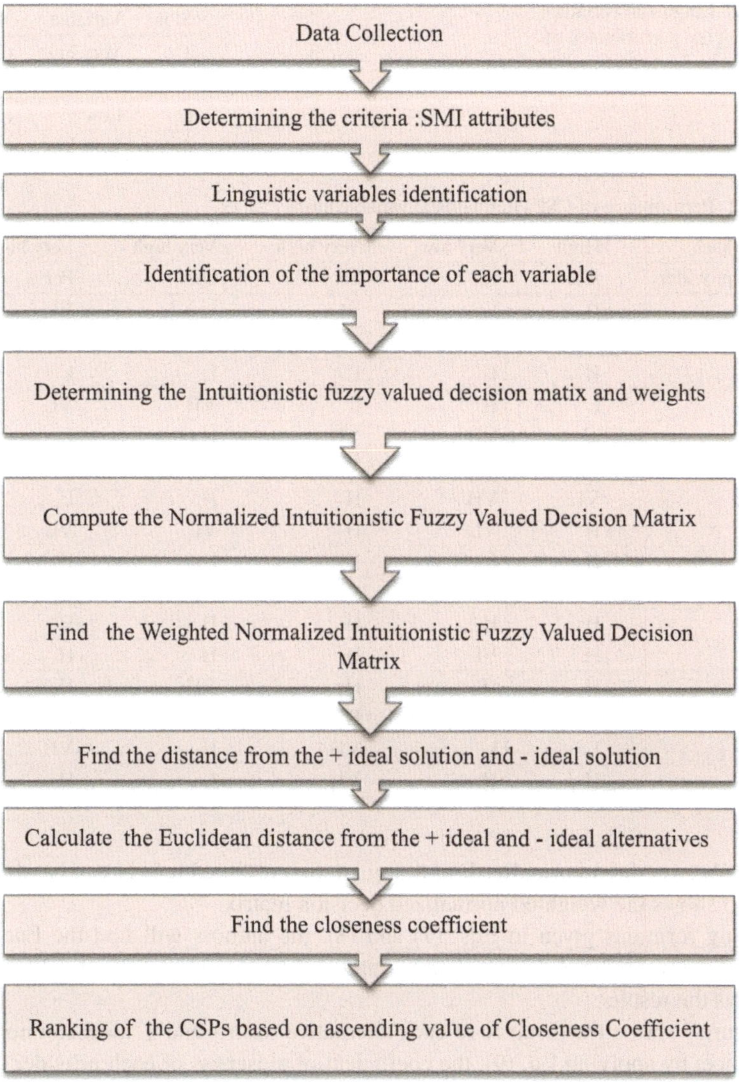

Fig. 2 Proposed ranking and selection approach

vidual cloud providers is gathered from the providers' websites and documentation. The linguistic variables for rating the performance of CSP and the significance of each criterion are shown in Table 1. An intuitionistic fuzzy value is associated with each linguistic variable.

In Table 2, the cloud service providers are indicated by CP1, CP2 etc. The columns acc, bil, sec, us and per are the criteria accountability, billing, security, usability and performance, respectively. By considering the importance of each

Table 1 Linguistic variables for rating the performance of CSP

Sl no	Variable	Value
1	Very high	(7,9,9)
2	Average	(3,5,7)
3	Low	(1,3,5)
4	Very low	(1,1,3)

Table 2 Performance of CSPs in linguistic variable terms

Importance	High	Very low	Very high	Very high	Very high
Cloud provider	Acc	Bil	Sec	Us	Per
CP1	H	L	VL	L	VL
CP2	L	L	VL	L	H
CP3	H	H	L	L	L
CP4	L	H	L	VH	VL
CP5	L	VH	VH	H	L
CP6	L	VL	VL	VL	VH
CP7	VL	VH	H	H	L
CP8	H	VL	H	VL	VL
CP9	H	L	VL	L	H
CP10	VL	L	VL	VH	L
CP11	H	H	H	H	H
CP12	H	VL	H	H	H
CP13	H	H	H	VH	H
CP14	VH	L	H	VH	H
CP15	L	L	VH	H	VH
CP16	H	H	VH	L	H

factor, the weighted normalized intuitionistic fuzzy valued matrix can be obtained. Table 3 shows the weighted normalized decision matrix.

Using formulas given in Eqs. (7) and (8), the authors will find the Euclidean distance between the alternatives and + and – ideal solutions, respectively. Table 4 displays the results.

Figure 3 shows the distance of each alternative from – and + ideal solutions.

Hence, by applying Eq. (9), the coefficient of closeness of each provider can be computed. Table 5 shows the results.

The graphical representation of the comparison of the closeness coefficients is shown in Fig. 4, as per the results shown above. The graph and table show that the coefficient of closeness of CSP 14 and CSP 13 is high and is followed by CSP 12 and CSP 11, and so on.

The final results are shown in Table 5. The results show that with regard to SMI criteria such as Accountability, Billing, Performance, Security and Usability, Providers 14 and 13 are the best among all available providers.

Figure 5 shows that the provider with the highest closeness coefficient has the first rank, and providers 14 and 13 are the best providers suitable for the consumer. The lowest rank is held by Provider 1.

Table 3 Fuzzy valued weighted normalized intuitionistic decision matrix

Weighted normalized fuzzy decision matrix															
Wt	5	7	9	1	1	3	7	9	9	7	9	9	7	9	9
CSP	Acc			Bill			Sec			Us			Per		
CP1	3.6	5.4	9.0	1.0	0.3	1.8	1.0	1.0	3.0	1.0	3.0	5.0	1.0	1.0	3.0
CP2	0.7	2.3	5.0	1.0	0.3	1.8	1.0	1.0	3.0	1.0	3.0	5.0	5.0	7.0	9.0
CP3	3.6	5.4	9.0	0.2	0.1	1.0	1.0	3.0	5.0	1.0	3.0	5.0	1.0	3.0	5.0
CP4	0.7	2.3	5.0	0.2	0.1	1.0	1.0	3.0	5.0	7.0	9.0	9.0	1.0	1.0	3.0
CP5	0.7	2.3	5.0	0.1	0.1	1.0	7.0	9.0	9.0	5.0	7.0	9.0	1.0	3.0	5.0
CP6	0.7	2.3	5.0	1.0	1.0	3.0	1.0	1.0	3.0	1.0	1.0	3.0	7.0	9.0	9.0
CP7	0.7	0.8	3.0	0.1	0.1	1.0	5.0	7.0	9.0	5.0	7.0	9.0	1.0	3.0	5.0
CP8	3.6	5.4	9.0	1.0	1.0	3.0	5.0	7.0	9.0	1.0	1.0	3.0	1.0	1.0	3.0
CP9	3.6	5.4	9.0	1.0	0.3	1.8	1.0	1.0	3.0	1.0	3.0	5.0	5.0	7.0	9.0
CP10	0.7	0.8	3.0	1.0	0.3	1.8	1.0	1.0	3.0	7.0	9.0	9.0	1.0	3.0	5.0
CP11	3.6	5.4	9.0	0.2	0.1	1.0	5.0	7.0	9.0	5.0	7.0	9.0	5.0	7.0	9.0
CP12	3.6	5.4	9.0	1.0	1.0	3.0	5.0	7.0	9.0	5.0	7.0	9.0	5.0	7.0	9.0
CP13	3.6	5.4	9.0	0.2	0.1	1.0	5.0	7.0	9.0	7.0	9.0	9.0	5.0	7.0	9.0
CP14	5.0	7.0	9.0	1.0	0.3	1.8	5.0	7.0	9.0	7.0	9.0	9.0	5.0	7.0	9.0
CP15	0.7	2.3	5.0	1.0	0.3	1.8	7.0	9.0	9.0	5.0	7.0	9.0	7.0	9.0	9.0
CP16	3.6	5.4	9.0	0.2	0.1	1.0	7.0	9.0	9.0	1.0	3.0	5.0	5.0	7.0	9.0
A*	5	7	9	1	1	3	7	9	9	7	9	9	7	9	9
A-	0.7	1	3	0	0	1	1	1	3	1	1	3	1	1	3

Table 4 Computed distance from the + ideal as well as – ideal solution

CSP	*dt+*	*dt–*
CP1	20.89	7.01
CP2	18.90	10.14
CP3	18.81	15.00
CP4	17.81	16.56
CP5	12.73	20.66
CP6	17.79	8.20
CP7	15.61	17.88
CP8	16.32	16.84
CP9	15.79	13.37
CP10	18.51	13.78
CP11	7.46	22.57
CP12	6.12	22.57
CP13	5.82	23.89
CP14	4.06	24.77
CP15	6.75	20.34
CP16	9.61	20.10

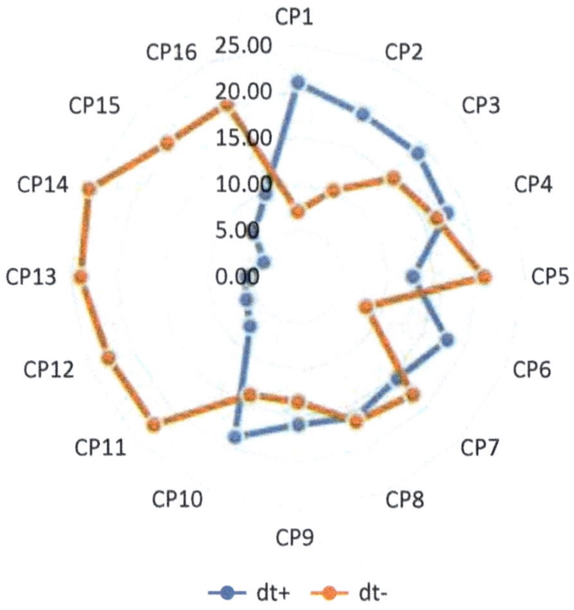

Fig. 3 Distance from + and − ideal solutions

Table 5 Coefficient of closeness of CSPs and ranks

CSP	CCi	Rank
CP1	0.25	16
CP2	0.35	14
CP3	0.44	12
CP4	0.48	10
CP5	0.62	7
CP6	0.32	15
CP7	0.53	8
CP8	0.51	9
CP9	0.46	11
CP10	0.43	13
CP11	0.75	4
CP12	0.79	3
CP13	0.80	2
CP14	0.86	1
CP15	0.75	5
CP16	0.68	6

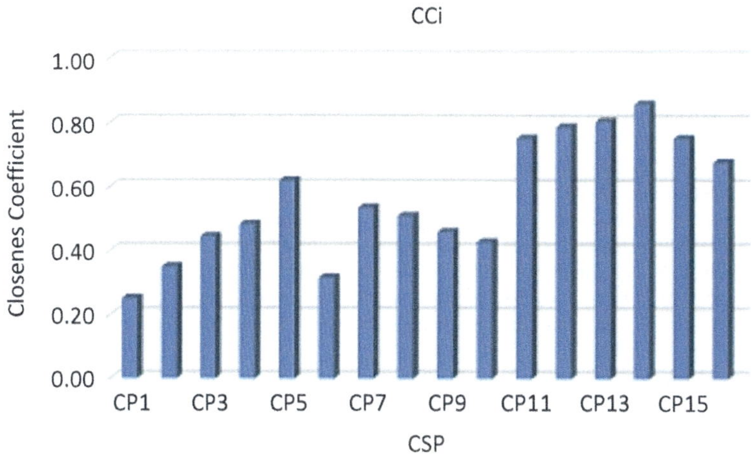

Fig. 4 CSPs and closeness coefficients

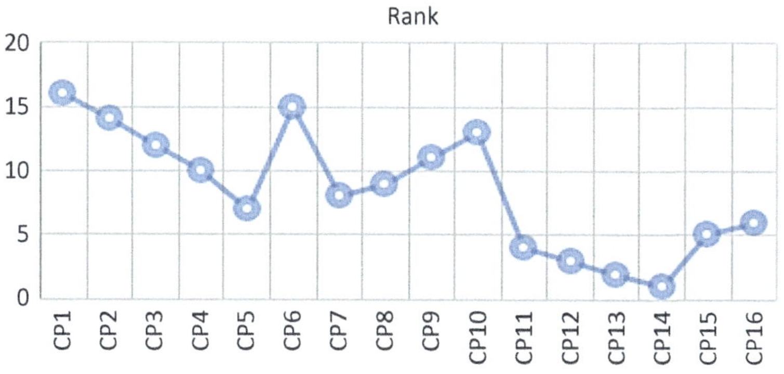

Fig. 5 CSPs and ranks

6 Conclusion

The choice of the best available Cloud Service Provider based on QoS attributes is a challenge for users especially. Qualitative and quantitative varieties fall under the selection criterion. The Cloud Service Provider selection based on SMI characteristics such as Usability, Billing, Accountability, Performance and Security was suggested in this chapter. The technique for order preference approach using the intuitionist fuzzy values is used for ranking the available providers based on the SMI attributes in this work. Using the respective linguistic variables, the ratings are interpreted and translated to intuitionist fuzzy values that efficiently rate the cloud providers. In the future, this research may be expanded to rate cloud services based on various MCDM approaches. By using new methods, fuzziness in selection can be managed effectively.

References

1. Siegel, J., & Perdue, J. (2012). Cloud services measures for global use: The service measurement index (SMI). In *2012 annual SRII global conference, San Jose, CA* (pp. 411–415).
2. Lee, S., & Seo, K.-K. (2016). A hybrid multi-criteria decision-making model for a cloud service selection problem using BSC, fuzzy Delphi method and fuzzy AHP. *Wireless Personal Communications, 86*(1), 57–75.
3. Tanoumand, N., et al. (2017). Selecting cloud computing service provider with fuzzy AHP. In *2017 IEEE international conference on fuzzy systems (FUZZ-IEEE)*. IEEE.
4. Chahal, R. K., & Singh, S. (2016). Fuzzy logic and AHP-based ranking of cloud service providers. In *Computational intelligence in data mining* (Vol. 1, pp. 337–346). Springer.
5. Roy, B. (1991). The outranking approach and the foundations of electre methods. *Theory and Decision, 31*(1), 49–73.
6. Tliga, H., & Rebaib, A. (2017). *A TOPSIS method based on intuitionistic fuzzy values: A case study of north African airports*. Growing Science. https://doi.org/10.5267/j.msl.
7. Ur Rehman, Z., Hussain, O. K., & Hussain, F. K. (2012). Iaas cloud selection using MCDM methods. In *2012 IEEE ninth international conference on e-business engineering (ICEBE)* (pp. 246–251). IEEE.
8. Garg, S. K., Versteeg, S., & Buyya, R. (2013). A framework for ranking of cloud computing services. *Future Generation Computer Systems, 29*(4), 1012–1023.
9. Peng, Y., Wang, G., & Wang, H. (2012). User preferences based software defect detection algorithms selection using MCDM. *Information Sciences, 191*, 3–13.
10. Fan, W. J., Yang, S. L., Perros, H., & Pei, J. (2015). A multi-dimensional trust-aware cloud service selection mechanism based on evidential reasoning approach. *International Journal of Automation and Computing, 12*(2), 208–219.
11. Kumar, R. R., Mishra, S., & Kumar, C. (2017). A novel framework for cloud service evaluation and selection using hybrid MCDM methods. *Arabian Journal for Science and Engineering, Springer, 43*, 7015–7030.
12. Baranwal, G., & Vidyarthi, D. P. (2014). A framework for selection of best cloud service provider using ranked voting method. In *2014 IEEE international advance computing conference (IACC)*. IEEE.

Similarity Analytics for Semantic Text Using Natural Language Processing

M. Karthiga, S. Sountharrajan, A. Bazila Banu, S. Sankarananth, E. Suganya, and B. Sathish Kumar

Abstract Determining the similarity among the sentences is a predominant task in natural language processing. The semantic determining task is one of the important research area in today's applications related to text analytics. The semantic of the sentences get varied according to the textual context it is used. In natural language processing, determining the semantic likeness between sentences is an important research area. As a result, a lot of research is done in determining the semantic likeness in the text. For example, there exists many possible semantics for a word (polysemy) and the synonym of the word; and also these techniques avoid considering the stop words in English which are critical for English phrase/word division, speech investigation, and meaningful comprehension. Our proposed work utilizes Term Frequency-based Inverse Document Frequency model and Glove algorithm-based word embeddings vector for determining the semantic similarity among the terms in the textual contents. Lemmatizer is utilized to reduce the terms to the most possible smallest lemmas. The outcomes demonstrate that the proposed methodology is more prominent than the TF-idf score in ranking the terms with respect to the search query terms. The Pearson correlation coefficient achieved for the semantic similarity model is 0.875.

M. Karthiga (✉) · A. Bazila Banu
Bannari Amman Institute of Technology, Coimbatore, Tamil Nadu, India
e-mail: karthigam@bitsathy.ac.in; bazilabanu@bitsathy.ac.in

S. Sountharrajan
VIT Bhopal University, Bhopal, Madhya Pradesh, India
e-mail: s.sountharrajan@vitbhopal.ac.in

S. Sankarananth
Anna University, Chennai, Tamil Nadu, India

E. Suganya
Excel College of Engineering and Technology, Komarapalayam, Tamil Nadu, India

B. Sathish Kumar
VIT University, Chennai, Tamilnadu, India
e-mail: sathiskumar.b@vit.ac.in

A. Haldorai et al. (eds.), *3rd EAI International Conference on Big Data Innovation for Sustainable Cognitive Computing*, EAI/Springer Innovations in Communication and Computing, https://doi.org/10.1007/978-3-030-78750-9_17

Keywords Term Frequency · Lemmatizer · Tokenizer · Semantic similarity · Word embeddings

1 Introduction

Semantic closeness detection is a quantifiable measure that illustrates the similarity in their meanings, when various sequences of text are provided without projecting on their nature of occurrence in the text. One of the major challenges in measuring semantic closeness in a natural language lies in the way of expressing the same information in a variety of manners when analysing the text context [1]. Thus, by utilizing the semantic closeness model, users are allowed to less repetitively reveal the same information meaningfully when evaluating the text messages thereby avoiding redundancy [2]. In the network-oriented applications, measuring semantic similarity among the terms is more useful for evaluating the search queries and for retrieving the most relevant textual contents [3]. In the proposed work, the term frequency in the textual contents with their similarity scores is achieved through Term Frequency-based Inverse Document Frequency model [4] and through Glove algorithm [5] in combination with word embeddings-based vector model. The model evaluates the textual contents with respect to the search query terms, and the similarity score is retrieved for the terms using the above said models. For different query terms, the similarity scores are retrieved, and the most relevant textual contents containing the terms with respect to the search query terms are retrieved, and the same could be utilized in search engines for retrieving the semantically relevant textual contents for the search query using the proposed work.

Current research in the area of Language Processing has given valuable results to semantic relations between the sentences and words [7, 8]. This part investigates the advantages and limitations of existing methods to identify the semantic measures [11]. It also strengthens the field of semantic web measures and relevance to co-relations methods [10]. Hanna Bechara et al. [20] proposed a semantic text similarity model using the classifier, support vector machine [12, 19]. Datasets are composed from the SemEval repository. The comparison is made between English and Spanish sentences. The system provides better results for English sentences compared to Spanish with an average ranking of 33 among 74 runs and a correlation result of 0.72 for Pearson using five different test datasets. Semantic similarity between two words using support vector machine is proposed by Karthiga et al. in [6, 9]. Semantic text closeness model for Hindi language using supervised learning approach and rule-based model was undergone by Darshan Agarwal [21]. For determining the semantic closeness among two sentences in Hindi language, Paninian reliance grammar that depicts the orientation of correlated words has been employed. It is the first system to measure semantic evaluation on Hindi sentences, and the test results provide about 79.9% correlation. A new semantic related measureness has been proposed in [15, 17] wherein wordnet features are taken into account.

Recent machine learning algorithms using natural language processing are utilized to extract actual models of similarities between the sentences and words. Word co-occurrence methods [13] are generally utilized in the Information Retrieval (IR) systems. It has a list of significant words, and every query is considered as a document. A vector is framed for query sets of documents. The related documents are extracted based on the likeness between the query vector and text vector [14]. Retrieving documents from multiple models and indexing them is a challenging task and the same is proposed in [16]. Image retrieval techniques based on the contents and the associated semantics in them are considered in [18].

2 Proposed Model

The best method of finding a relevant document is through searching. Searching always retrieves some documents according to the keywords given in search query. But, the documents retrieved may or may not be meaningful with respect to the search query. If the documents are retrieved with respect to the meanings concerned with the search query, then most relevant documents would be retrieved. This could be possible through various Natural Language Processing methods. In the proposed method, Term Frequency (TF)–Inverse-Document-Frequency (idf) method and word embeddings-based semantic similarity measure are utilized, and both the methods are compared against the dataset utilized. In TF-idf method, the words that occur in different text contents are ranked according to the number of times they occur in the textual content. It is one of the most elegant tools in places where occurrence of same words and necessity of words in the textual content play an important role. In semantic similarity model, the words are ranked in accordance with their similarity and not with their match. It utilizes NLP techniques such as word embeddings to rank the words when words that overlap in the textual content are small.

Initially in the proposed work, TF-idf model alone is used, and the results are gathered. Then, semantic similarity measure is used as it requires lot of functions from Gensim library and Glove algorithm. The procedure of the proposed work is represented in Fig. 1.

The proposed model utilizes a lexical database [21] to determine the appropriate semantic of the word in comparison with the other existing techniques that utilize a fixed vocabulary. For every textual content, a semantic vector is created that contains the rank of the words in terms of weight values in comparison with the words in the second textual content used for comparison. Along with word frequency, information is also considered. The steps involved in the proposed model are represented below:

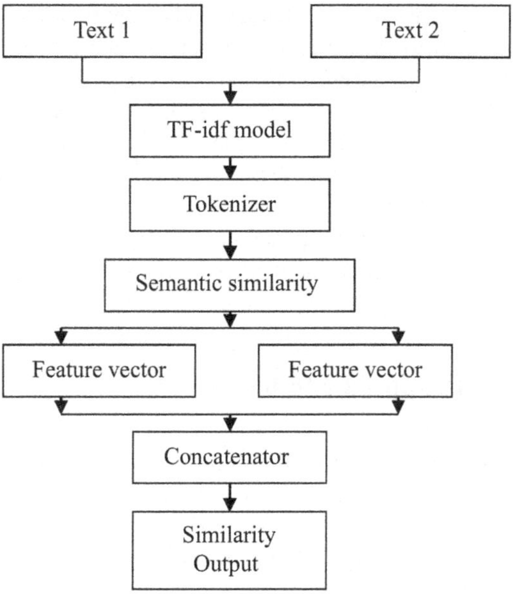

Fig. 1 Proposed prototype structural representation

2.1 *TF-idf*

This is a significant method for ranking similarity of the documents depending on the importance of the words present in the document. If a term (word) occurs more than once in a document, then that word is important for that document. Also, if a word (term) occurs number of times in many documents, then that word (term) is not so important. The score produced by TF-idf is based on considering the above two ideas of the frequency of occurrence of a word (term). TF-idf score helps to determine the importance of a term (word) in a textual content. This weighting methodology is utilized by many searching engines for ranking documents with respect to the search query. This is also utilized in stop words filtering in many applications. TF-idf is calculated based on two steps: first is Term Frequency (TF) and the second is Inverse Document Frequency (idf). TF is measured as the ratio of occurrence of term (word) in a textual content to the entire number of terms (words) in the textual content. TF also measures occurrence of a term in a textual content. Considering the length of the textual content as a term appears many times in a long textual content and small number of times in a small document, TF is always divided by the length of the textual content. Equation (1) represents the formula for determining the Term Frequency (TF):

$$\mathrm{TF} = \frac{\text{Number of times the term (word) w appears in a textual content}}{\text{Total number of words (terms) in a textual content}} \tag{1}$$

Inverse Document frequency (idf) is computed as logarithmic ratio of the entire number of textual contents to number of textual contents with that word (term) in it. It computes how significant a word (term) is to a textual content. While measuring the TF, all the words (terms) in a textual content are considered evenly. But some words like "is", "that", "are", "of", "the" appears many times in a textual content but they hold less importance. Thus, frequent words (terms) need to be weighted low than the less frequent terms (words) in a document. The idf is measured using the Eq. (2):

$$\text{idf} = \log_e \frac{\text{Total number of textual contents}}{\text{Number of documents containing the word (term) in it}} \tag{2}$$

TF-idf vectorization model is imported from scikit learn package that converts group of raw textual contents to matrix of TF-idf features. The preprocessing step of tokenization is performed by the TFidfVectorizer where each single word (term) is changed to tokens. It also creates term (word) frequencies containing sparse matrix. Then, dot product that is linear kernel of the initial vector with the textual contents is computed to find the similarity. The initial vector is the textual content containing the search words (terms). The first result that is ([1:1]) is ignored because it is comparing search words to itself. Initially, the stop words are ignored as they do not make good sense in the textual corpus while computing similarity. Also as the TF-idf model simply determines the exact matches and the words "flower" and "flowers" are not considered as same, another approach named lemmatizer is utilized to break down each word (term) to its reduced lemma. Using lemmatizer, the term "pomegranates" are reduced to "pomegranate" and after reduction, these terms are compared to its related terms in the textual content.

In a dimensional space, each textual document is a point wherein the total number of dimensions in that dimensional space relates to the number of unique terms in that textual content. This dimensional space becomes empty when the distance between the points in that dimensionality is more, which means the chosen textual content contains more number of words (terms) in it. In such cases, the angle between the document points is considered for determining the similarity. Some distance metrics like Manhattan distance, Euclidean distance could be used to determine the distance between the document points. In the proposed method, cosine similarity is utilized as the output from TF_idfVectorizer is L2-normalized.

2.2 Semantic Similarity

Another approach utilized in the proposed work to determine the similarity is through semantic similarity that compares the textual contents based upon the similarity among the terms in the textual content. The terms in the textual contents are ranked according to the togetherness they occur in the textual content. This model is known as Glove model, and it is one of the word embeddings techniques. In Glove model, the terms in the textual contents are mapped to the vectors that are creating points in the multidimensional space. If the points distance in the dimensional space is less, then the terms in the textual content occur together. Glove algorithm is an unsupervised learning method created in 2014 at Stanford University. For preprocessing, the terms, that is, conversion of terms to tokens, Gensim library is utilized. This library along with tokenization facilitates the removal of HTML tags if the textual contents are extracted from the web. After that a similarity matrix is formed that consists of similarity among each term pairs. The similarities are weighted depending on the frequency of appearance of terms in the textual content. Finally, soft-cosine similarity is measured among the search query and the textual contents. The soft-cosine similarity is better than normal cosine similarity as it takes into account the term similarity also.

3 Results and Analysis

The algorithm is evaluated using the standard dataset that contains 68 pairs calculated by Rubenstein et al. in [22]. This data is one of the widely utilized dataset for many applications related to semantic similarity. The model dataset taken is represented in Fig. 2.

The dataset is loaded into the working environment. Then TF-idfVectorizer is imported from scikit library package for tokenization of the textual contents. Linear model is imported from scikit learn metrics library for computing the similarity. Also stop words package is imported from nltk.corpus package. After installing and downloading all the required packages into the working environment, vectors are formed between the search terms and the textual contents using TF-idfVectorizer. Term frequency and cosine similarity are computed for the search terms and the


```
{ "data": [["Pomegranate Bhagwa","Fresh Pomegranate from Anushka Avni International Bhagwa is a premium Pomegranate variety from India. The Deep Red arils & the pleasing Red b
["Pomegranate Arakta","Fresh Pomegranate Arakta from Anushka Avni International This Pomegranate are bigger in size, sweet with soft seeds, bold red arils. It also possess glo
["About Us","About Us Anushka Avni International (AAI) takes pleasure in presenting itself as one of the renowned Suppliers and Exporter. We have huge assortment of agro produ
["Contact Us","About Us Anushka Avni International (AAI) takes pleasure in presenting itself as one of the renowned Suppliers and Exporter. We have huge assortment of agro pro
["White Onions","White Onions from Anushka Avni International Fresh White Onion, which is widely acclaimed for its health benefits and is used for different cuisines. We are a
["Video Gallery","Anushka Avni International (AAI) takes pleasure in presenting itself as one of the renowned Suppliers and Exporter. We have huge assortment of agro products
["Major Market","UK, Nether Land, Russia, Canada, HongKong, Malaysia, Singapore, Oman, UAE, Qatar, Bahrain, Saudi Arabia, Mauritius, Maldives, Sri Lanka, Bangladesh"],
["Red Onions","Red Onions from Anushka Avni International Red onions, sometimes called purple onions, are cultivars of the onion with purplish red skin and white flesh tinged
["Grapes Sonaka Seedless","Berry Size: 18mm and above Packaging Packing Size: 4.5 kg loose in carry bags 8.2 kg loose in carry bags 5 kg carton punnets10 x 500 gms Load abilit
["Pink Onions","Pink Onions from Anushka Avni International Pink onions, sometimes called summer onions, are cultivars of the onion with pink skin and white flesh tinged with
["About Us","Anushka Avni International (AAI) takes pleasure in presenting itself as one of the renowned Suppliers and Exporter. We have huge assortment of agro products avail
["Small Onions","We are one of the leading organizations engaged in delivering our customers with Fresh Onions. We manufacture this in bulk requirements for our clients. Malay
["Tomatoes","Fresh Tomatoes from Anushka Avni International We have emerged as one of the reputed organization actively participating in exporting and suppling Red Tomatoes. P
["Downloads","Anushka Avni International (AAI) takes pleasure in presenting itself as one of the renowned Suppliers and Exporter. We have huge assortment of agro products avai
["Welcome to Anushka Avni International (AAI)","Welcome to Anushka Avni International (AAI) Anushka Avni International (AAI) takes pleasure in presenting itself as one of the
["Grapes Thompson Seedless","Thomson Seedless These grapes are seedless, sweet-tart, and crunchy. Thomson Seedless account for the bulk of Table Grape exports from India. Berr
["Grapes Black Sharad Seedless","Black Sharad Seedless Grapes Crisp to taste, these grapes vary in shades of black to purple. Each berry is usually medium-sized and oval-shape
["Grapes Flame / Red Seedless","Flame / Red seedless grapesFlame grapes are one of the most popular varieties along with…"],
["DA-TEX sewing machine","Power Requirements Ac/Dc Motor: 1/10 Hp 8000 Rpm 220v , 1.5 Amps Also Available With 110 v AC/DC Motor Sewing Speed 1600 ? 1800 Stitches Per Minute 5
["LPI DA ? 1 Extra Heavy Duty","Power Requirements Ac/Dc Motor: 1/10 Hp 8000 Rpm 220v , 1.5 Amps Also Available With 110 v AC/DC Motor Sewing Speed 1600 ? 1800 Stitches Per Mi
["WAZIR DE-DA sewing machine","New Model : DE-DA Power:- AC/DC MOTOR: 1/10 Requirement:- 1/10 hp 8000rpm 220V, 1.5 amps Also available with 110V AC/DC Motor. : Sewing Speed :-
["LPI DE-DA sewing machine","New Model : DE-DA Power:- AC/DC MOTOR: 1/10 Requirement:- 1/10 hp 8000rpm 220V, 1.5 amps Also available with 110V AC/DC Motor. : : Sewing Speed :-
["LPI DA-6 sewing machine","Features :- Power Requirements Ac/Dc Motor: 1/10 Hp 8000 Rpm 220v , 1.5 Amps Also Available With 110 v AC/DC Motor Sewing Speed 1600 ? 1800 Stitche
["LPI DA sewing machinge","New Model : DA Power:- AC/DC MOTOR: 1/10 Requirement:- 1/10 hp 8000rpm 220V, 1.5 amps Also available with 110V AC/DC Motor. : Sewing Speed :-1600 -1
["How to put Thread in Bag Closer LPI","How to put Thread in Bag Closer LPI       <!-- !function(a,b){function c(){if(!e){e=!0;var a,c,d,f,g=-1!==navigator.appVersion.indexOf(
```


Fig. 2 Sample dataset

textual contents using linear_kernel.flatten() function. The top-scoring results are retrieved for various search terms, and the same is represented in Table 1.

From Table 1, it is depicted that for the search terms 'fruits and vegetables', the number of textual contents that score non-zero values are two as they contain the word 'fruit' in the textual content. If the search term is 'onion', there are no matches in the textual contents as only the plural 'onions' are there. Since 'onion' and 'onions' are both same, the above TF-idf model is unable to recognize the match and to overcome the above problem, lemmatizer is used along with TF-idf for reducing the terms to its shorter form to determine the exact match. This lemmatizer is very much useful while plurals are there in the textual contents. For utilizing that, wordnetlemmatizer package is imported from nltk package. The similarity outputs after utilizing TF-idf along with lemmatizer is represented in Table 2.

From Table 2, it is understood that for the search term 'onions', the textual content "White Onions" is scoring high than other textual contents. For semantic similarity matching using Glove embeddings, Gensim library is installed along with all necessary packages in the working environment. After downloading and loading the Glove word vector embeddings, dictionary for terms in the textual content is built using TF-idf model. The search terms also should be in the built dictionary. Term similarity matrix is computed. Using the similarity matrix, the soft-cosine similarity score is evaluated among the search term and the textual contents. Finally, the output of the textual content semantic similarity model by utilizing the Glove word embeddings is represented in Table 3. For each of the term present in the search query, the most similar terms in the textual content is extracted, and the results of the same are depicted in Table 4.

Table 4 reveals that the terms in the textual contents are most similar to the search query terms. Thus, the dataset is used for two models in the proposed work, and for our dataset containing less query search terms, semantic similarity model with Glove word embeddings showed high similarity scores compared to the TF-idf model. The comparison results are depicted in Fig. 3. From Fig. 3, it is revealed that semantic similarity score achieves best results compared to the TF-idf score for the chosen dataset. The Pearson correlation coefficient achieved for the TF-idf model is 0.825 whereas for Glove word-embeddings model is 0.875.

Table 1 Similarity score based on TF-idf

Search terms		Fruit and vegetables
S. no	Text contents	Similarity score
1	Pomegranate Bhagwa	0.122
2	Pomegranate Arakta	0.046
3	About Us	0.000
4	Contact Us	0.000
5	White Onions	0.000

Table 2 Similarity score based on TF-idf with lemmatizer

Search terms		Onions
S. no	Text contents	Similarity score
1	White Onions	0.365
2	Pomegranate Arakta	0.000
3	About Us	0.000
4	Contact Us	0.000

Table 3 Textual contents similarity score

Document index	Similarity score	Text contents
0	0.695	Pomegranate Bhagwa
11	0.631	Small Onions
12	0.615	Tomatoes
16	0.590	Grapes Black Sharad Seedless
1	0.570	Pomegranate Arakta
4	0.523	White Onions
10	0.495	About Us
7	0.480	Red Onions
9	0.471	Pink Onions
19	0.462	LPI DA? 1 Extra Heavy Duty
18	0.428	DA-TEX sewing machine

Table 4 High relevant terms in the textual contents for the search terms

Document index	Similarity score	Text contents	Search terms
0	0.695	Pomegranate Bhagwa	Fruit, delicious, sweet, cherry, fresh
11	0.631	Small Onions	Fresh
12	0.615	Tomatoes	Fresh, taste, tomatoes
16	0.590	Grapes Black Sharad Seedless	Grapes, taste
1	0.570	Pomegranate Arakta	Fruit, sweet, fresh, taste, soft
4	0.523	White Onions	Fresh
10	0.495	About Us	Harvest
7	0.480	Red Onions	Fresh, flesh
9	0.471	Pink Onions	Fresh, flesh
19	0.462	LPI DA? 1 Extra Heavy Duty	Cotton
18	0.428	DA-TEX sewing machine	Cotton

4 Conclusion

This chapter reveals about calculating the semantic similarity between the terms in the textual contents. The model initially tokenized each of the terms to individual tokens followed by reducing the terms to its smallest lemma using lemmatizers. Then the model utilizes Term Frequency Inverse Document Frequency approach

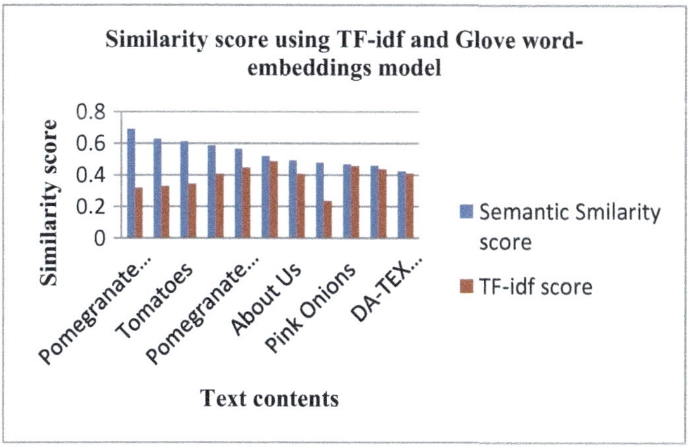

Fig. 3 Similarity score using TF-idf and Glove word-embeddings model

to compute the similarity among the terms in the textual contents. The results are built as a form of dictionary, and the same is utilized for the semantic similarity score using Glove algorithm in combination with word-embeddings vector. The semantic similarity score is computed for each search query term, and the results are compared with the TF-idf scores. The Pearson correlation coefficient obtained for the proposed model is 0.875, which is better than 0.825 score obtained with TF-idf.

References

1. Lecun, Y., & Bengio, Y. (1995). Convolutional networks for images, speech, and time series [J]. *The Handbook of Brain Theory and Neural Networks, 3361*(10), 1995.
2. Huang, P.-S., He, X., Gao, J., et al. (2013). Learning deep structured semantic models for web search using click through data[C]. In *Proceedings of the 22nd ACM international conference on information and knowledge management, Amazon, India* (pp. 2333–2338).
3. Shen, Y., He, X., Gao, J., et al. (2014). A latent semantic model with convolutional-pooling structure for information retrieval[C]. In *Proceedings of the 23rd ACM international conference on information and knowledge management, New York, USA* (pp. 101–110).
4. Palangi, H., Deng, L., Shen, Y., et al. (2016). Deep sentence embedding using long short-term memory networks: Analysis and application to information retrieval [J]. *IEEE/ACM Transactions on Audio, Speech, and Language Processing, 24*(4), 694–707.
5. Sountharrajan, S., et al. (2020). Dynamic recognition of phishing URLs using deep learning techniques. In *Advances in cyber security analytics and decision systems* (pp. 27–56). Cham: Springer.
6. Karthiga, M., & Priyadharsini, M. (2013). A semantic search engine using semantic similarity measure between words. *International Journal of Scientific and Engineering Research, 4*(5), 379–384.

7. Wan, S., Lan, Y., Guo, J., et al. (2016). A deep architecture for semantic matching with multiple positional sentence representations[C]. In *Proceedings of the 30th AAAI conference on artificial intelligence, Phoenix, USA* (pp. 2835–2841).
8. Yuzhi, S. (2015). *Chinese grammar*. Beijing: The Commercial Press. (in Chinese).
9. Karthiga, M., Kalaivaani, P. C., & Sankarananth, S. (2013, February 2). A semantic similarity approach based on web resources. In *2013 international conference on information communication and embedded systems (ICICES)* (pp. 226–231). IEEE.
10. Abney, S., & Light, M. (1999). Hiding a semantic class hierarchy in a Markov model. In *Proceedings of the ACL workshop unsupervised learning in natural language processing* (pp. 1–8).
11. Budanitsky, A., & Hirst, G. (2001, June). Semantic distance in WordNet: An experimental, application-oriented evaluation of five measures. In *Proceedings of the workshop WordNet and other lexical resources, second meeting north American chapter of the Association for Computational Linguistics*.
12. Suganya, E., et al. (2019). Mobile cancer prophecy system to assist patients: Big data analysis and design. *Journal of Computational and Theoretical Nanoscience, 16*(8), 3623–3628.
13. Otegi, A., & Ansa, A. (2015). Using knowledge-based relatedness for information retrieval. *Knowledge and Information Systems, 44*(3), 689–718.
14. Liu, Q., Liu, B., Zhang, Y., Kim, D. S., & Gao, Z. Q. (2016). Improving opinion aspect extraction using semantic similarity and aspect associations. In *Proceedings of the 30th AAAI conference on artificial intelligence Phoenix* (pp. 2986–2992).
15. Taieb, M. A. H., Aouicha, M. B., & Hamadou, A. B. (2014). A new semantic relatedness measurement using WordNet features. *Knowledge and Information Systems, 41*(2), 467–497.
16. Srihari, R. K., Zhang, Z. F., & Rao, A. B. (2000). Intelligent indexing and semantic retrieval of multimodal documents. *Information Retrieval, 2*, 245–275.
17. Miller, G. A., Bechwith, R., Felbaum, C., Gross, D., & Miller, K. (1990). Introduction to WordNet: An on-line lexical database. *International Journal of Lexicography, 3*(4), 235–244.
18. Smeulders, Worring, M., Santini, S., Gupta, A., & Jain. (2000). Content-based image retrieval at the end of the early years. *IEEE Transactions on Pattern Analysis and Machine Intelligence, 22*(12), 1349–1380.
19. Bechara, H., et al. (2015, June 4–5). MiniExperts: An SVM approach for measuring semantic textual similarity. In *Proceedings of the 9th international workshop on Semantic Evaluation (SemEval 2015), Denver, Colorado* (pp. 96–101). Association for Computational Linguistics.
20. Agarwal, D., et al. (2017). Semantic textual similarity For Hindi. In *Proceedings of 18th International Conference on Computational Linguistics and Intelligent Text Processing (CICLing-2017), Budapest, Hungary.*
21. Miller, G. A. (1995). WordNet: A lexical database for English. *Communications of the ACM, 38*(11), 39–41.
22. Rubenstein, H., & Goodenough, J. B. (1965). Contextual correlates of synonymy. In *Communications of the ACM 8.10* (pp. 627–633).

Index

A. Haldorai et al. (eds.), *3rd EAI International Conference on Big Data Innovation for Sustainable Cognitive Computing*, EAI/Springer Innovations in Communication and Computing, https://doi.org/10.1007/978-3-030-78750-9

The manufacturer's authorised representative in the EU is Springer Nature Customer Service Centre GmbH, Europaplatz 3, 69115 Heidelberg, Germany. If you have any concerns regarding our products, please contact ProductSafety@springernature.com

Printed and bound by CPI Group (UK) Ltd, Croydon, CR0 4YY
30/09/2026
02225291-0005